SpringerBriefs in Computer Science

Series Editors

Stan Zdonik
Shashi Shekhar
Jonathan Katz
Xindong Wu
Lakhmi C. Jain
David Padua
Xuemin (Sherman) Shen
Borko Furht
V.S. Subrahmanian
Martial Hebert
Katsushi Ikeuchi
Bruno Siciliano
Sushil Jajodia
Newton Lee

More information about this series at http://www.springer.com/series/10028

Yingjiu Li • Qiang Yan • Robert H. Deng

Leakage Resilient Password Systems

 Springer

Yingjiu Li
School of Information Systems
Singapore Management University
Singapore

Qiang Yan
Google, Zurich
Switzerland

Robert H. Deng
School of Information Systems
Singapore Management University
Singapore

ISSN 2191-5768   ISSN 2191-5776   (electronic)
SpringerBriefs in Computer Science
ISBN 978-3-319-17502-7   ISBN 978-3-319-17503-4   (eBook)
DOI 10.1007/978-3-319-17503-4

Library of Congress Control Number: 2015936731

Springer Cham Heidelberg New York Dordrecht London

Printed on acid-free paper

Springer International Publishing AG Switzerland is part of Springer Science+Business Media (www.
springer.com)

Preface

The design of leakage resilient password (LRP) systems in the absence of any trusted devices for unaided users remains a challenging problem despite two decades of intensive research in the security community. The first chapter of this book investigates the inherent tradeoff between security and usability in designing LRP systems. It is demonstrated that most of the existing LRP systems are subject to two types of generic attacks, the brute force attacks and the statistical attacks, and that these attacks cannot be effectively mitigated without sacrificing the usability of LRP systems. A quantitative analysis framework is introduced on the usability of LRP systems for which the authentication process is decomposed into some atomic cognitive operations in psychology. It is concluded that a secure LRP system in practical settings imposes a considerable amount of cognitive workload unless certain secure channels are involved.

The second chapter of this book introduces a secure and practical LRP system, named CoverPad, for password entry on touch-screen mobile devices. CoverPad leverages a temporary secure channel between a user and a touch screen, which can be easily realized by placing a hand-shielding gesture on the touch screen. The temporary secure channel is used to deliver a hidden message to the user for transforming each password symbol before entering it on the touch screen. CoverPad is proven to be leakage resilient and it retains most of the benefits of legacy passwords. The usability of CoverPad is evaluated in a rigorous user study with realistic testing conditions including time pressure, distraction, and mental workload. The user study shows the impact of these testing conditions on the users' performance in practice.

The third chapter of this book introduces a new LRP system, named ShadowKey. ShadowKey is designed to achieve better usability for leakage resilient password entry. ShadowKey leverages either permanent secure channel, which naturally exists between a user and the display unit of certain mobile device such as smart glasses, or temporary secure channel, which can be easily realized between a user and a touch screen with a hand-shielding gesture. The secure channel is used to protect the mappings between original password symbols and associated random symbols. After seeing the mappings, the user of ShadowKey can freely input the random symbols, instead of the original password, via a voice-input channel or on a touch screen under no protection. ShadowKey is easy to use in a sense that users do not need to remember anything else except their passwords, and they do not need to perform any transformation operations in their minds as it is required in the previous LRP systems.

Focus and Audience of This Book

User authentication is the first line of defence for securing most computing systems and services. Password based user authentication remains the most pervasive in user authentication practice due to many benefits including high usability and low cost. When applied on mobile devices, especially when used in public places, password based user authentication incurs a major concern on password leakage. This book focuses on the design, analysis, and evaluation of leakage resilient password systems. This focus is different from, and complementary to, many other publications on password systems, which are not designed to be leakage resilient.

The intended audience of this book includes academic researchers and graduate students who are interested in exploring the recent development of leakage resilient password systems. They may find it useful in understanding the major challenges in designing leakage resilient password systems, and inspiring new ideas from existing solutions. This book is also suitable for password system engineers and practitioners to design practical leakage resilient password systems. They may find available state-of-the-art solutions, which can be implemented on a wide variety of devices such as smart glasses, smart phones, laptops, and touch-screen desk-tops and ATMs for users to input their passwords/PINs easily in public places without leaking out any information about their passwords. The solutions provided in this book may serve as references, and thus should be interesting to mobile device manufacturers, smart glasses manufacturers, and user authentication software and service providers.

Acknowledgements

Some of the material in this book has appeared elsewhere. The authors gratefully acknowledge the copyright holders, including the Internet Society for permission to use material from "On Limitations of Designing Usable Leakage-Resilient Password Systems: Attacks, Principles and Usability," *Proceedings of the 19th Network & Distributed System Security Symposium (NDSS)*, 2012 in Chap. 1, and the Association for Computing Machinery, Inc. for permission to use material from "Designing Leakage-Resilient Password Entry on Touchscreen Mobile Devices," *Proceedings of the 8th ACM SIGSAC Symposium on Information, Computer and Communications Security*, pp. 37–48, 2013 in Chap. 2. The authors also acknowledge Elsevier for permission to use material from "Leakage-Resilient Password Entry: Challenges, Design, and Evaluation," *Computers & Security*, Vol. 48, pp. 196–211, 2015 in Chap. 2.

The authors would like to thank their collaborators Dr. Jin Han and Dr. Jianying Zhou for valuable contributions in the research and discussions.

The work that is led to the material published in this book also provides the research foundations for the project of secure and usable authentication systems in mobile computing that is part of the Secure Mobile Centre at Singapore Management University, which is supported by the Singapore National Research Foundation under NCR Award Number NRF2014NCR-NCR001-012.

Singapore   Yingjiu Li

Zurich, Switzerland   Qiang Yan

Singapore   Robert H. Deng

January 2015

Contents

Chapter 1
Leakage Resilient Password Systems: Attacks, Principles, and Usability

1.1 Introduction

A password (or PIN, pass phrase, pass code, access code) is a string of symbols used for user authentication. Passwords have been commonly used for user authentication in a wide variety of electronic systems, including computers, mobile devices, web services, access control systems, and ATM. Compared to PKI, biometrics, and smart cards which could be used for user authentication, passwords are much easier to create, update, and revoke without relying on any additional hardware or underlying infrastructure. Not surprisingly, the legacy password systems remain the most pervasive in user authentication due to their advantages of high usability and low costs.

One of the most common security threats to password systems is the leakage of passwords [21]. In a password leakage attack, an adversary may steal a password by capturing (e.g. via shoulder-surfing, video taping or key logging) and analyzing a user's inputs during password entry. Such threat is particularly relevant to mobile devices since they are often used in public places. Conventional password systems require that a user directly input his/her entire plaintext password which is recalled from the user's memory. In such case, an observation of a single authentication session is sufficient for an adversary to capture the password.

In order to prevent password leakage during password entry, a user should not input the original password but a transformed one, while the transformation of the password may impose an extra burden on the user. A fundamental problem in the design of leakage resilient password (LRP) system is how to minimize the password leakage and still make it easy to use.

An ideal LRP system allows a user to generate a one-time password (OTP) for each authentication session based on the original password such that an adversary cannot infer any useful information about the original password from the observed OTPs. In some cases, an LRP system allows a user to use a secure channel when interacting with the system.

© The Author(s) 2015
Y. Li et al., *Leakage Resilient Password Systems*, SpringerBriefs
in Computer Science, DOI 10.1007/978-3-319-17503-4_1

In Chap. 1 of this book, we investigate the feasibility of designing a usable and secure LRP system without using any secure channels [34]. In Chaps. 2 and 3, we explore certain easy-to-establish temporary secure channels for the design of usable and secure LRP systems. Also in Chap. 3, we explore certain existing permanent secure channels to further improve the usability of LRP systems. Note that in this book, we do not consider the secure channels which are formed using additional trusted devices such as secure tokens and smart cards because they require extra hardware, and they are subject to theft and loss.

1.2   Models

In this chapter, we focus on the fundamental problem of designing LRP systems for unaided humans using no secure channels during password entry. An LRP system is essentially a challenge-response protocol between human and computer. We refer to human as *user*, and computer as *server*. During registration, a user and a server agree on a *root secret*, usually referred to as a password. The user later uses the root secret to generate *responses* to *challenges* issued by the server to prove his/her identity. Unlike conventional password systems, a response in an LRP system is an obfuscated message derived from the root secret, rather than the plaintext of the root secret itself. Considering the limited cognitive capabilities of unaided humans, a usable obfuscation function F is usually a many-to-one mapping from a large candidate set to a small answer set. The small size of the answer set increases the success rate of a *guessing attack* where an adversary attempts to pass the authentication by randomly picking an answer from the answer set. For this reason, an *authentication session* of an LRP system often requires executing multiple rounds of the challenge-response procedure in order to reach an expected authentication strength D. D can be defined as the resistance against random guessing. For example, $D = 10^{-6}$ for 6-digit PIN. Each round of the challenge-response procedure is referred to as an *authentication round*. We use d to denote the average success rate of the guessing attack per authentication round. Given d and D, the minimum number m of authentication rounds for an authentication session is $\lceil \log_d D \rceil$.

A *k-out-of-n paradigm* [15] is used in many existing LRP systems [2, 15, 20, 26, 30, 31, 35]. In this paradigm, the root secret consists of k independent elements randomly drawn from a pool of n elements. An element can be an image, a text character, or any symbol in a notational scheme. The set of k secret elements is called the *secret set*, which forms the root secret of the user, and the complementary set is called the *decoy set*. The server knows the secret set chosen by the user, and uses a subset or all of these k elements in generating a challenge in each round. We refer to the chosen portion of the root secret for an authentication round as *round secret*.

A tuple (D, k, n, d, w, s) is used to describe the common parameters for the most existing LRP systems [2, 4, 15, 20, 26, 30, 31, 35], where D is the authentication strength, k is the number of secret elements drawn from an alphabet of n candidate elements, d is the average success rate of the guessing attack in a single round, w is the average window size (i.e., the number of elements shown to the user in each round), and s is the average length of a user's decision path (i.e., the number of decisions a user must make before producing the correct response for each round). The total number m of rounds can be derived from D and d.

Fig. 1.1 Setting of LRP system in Chap. 1

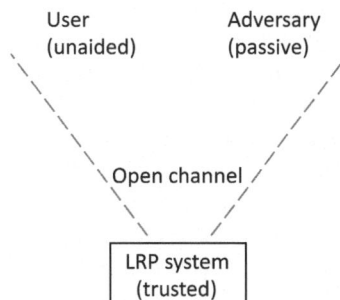

1.2.1   Threat Model

Figure 1.1 shows the setting of LRP system in Chap. 1, where the LRP system is trusted, and the channel between the LRP system (i.e., server) and an unaided user is open to passive adversaries. Two types of passive adversaries have been considered in the prior research on LRP systems. A weak passive adversary model (e.g. *cognitive shoulder-surfing* [25]) assumes that an adversary is not able to capture the complete interaction between user and server [25]. This assumption, however, may not hold for a prepared adversary who deploys a *hidden camera, key logger, or phishing web site* to capture the whole password entry process. To address such realistic concerns, recent research on LRP systems [2, 4, 20, 26, 30, 31, 35] takes a strong passive adversary model, where an adversary without any prior knowledge is allowed to record the complete interaction between user and server. We continue to use such strong passive adversary model in this book.

In the strong passive adversary model, the password leakage may happen during human-computer interactions if a user's inputs can be distinguished from random messages. After recording a sufficient number of authentication rounds, an adversary may use any reasonable computation resources to analyze and recover the underlying password from the user's inputs. The research problem in this case is to lower the leakage rate while maintaining an acceptable level of usability in the design of LRP systems.

Although many LRP systems have been proposed under the strong passive adversary model [2, 4, 15, 20, 26, 30, 31, 35], we will show that all of the proposals with acceptable usability are vulnerable to one or two generic types of attacks, the brute force attacks and the statistical attacks.

The brute force attacks can be considered as a pruning process for the entire candidate password set. We will show that the brute force attacks can be used to recover the root secrets of certain existing LRP systems from a small number of observations of authentication sessions. The statistical attacks, on the other hand, represent a learning process of extracting a user's secret according to the statistical significance of the secret. Two types of statistical attacks, including probabilistic decision tree and multi-dimensional counting, are investigated to show the vulnerability of existing LRP systems.

We note that these two generic types of attacks are different from the specific attacks of SAT [13] and Gaussian elimination [19] which have been systematically studied in the literature. The SAT attacks can be effectively mitigated by requiring a user to select only one correct response when multiple correct responses can be derived for each challenge. On the other hand, the Gaussian elimination-based algebraic attacks can be prevented by using a non-linear response function [20] or introducing certain noises such as intentional mistakes [15]. Unlike these specific attacks, the brute force attacks and the statistical attacks cannot be easily defended without significantly sacrificing the system's usability, which reveals an inherent limitation of LRP systems.

The security strength of an LRP system is defined as the resistance against these two generic types attacks which is measured as the equivalent success rate of random guessing. The security strength of existing LRP systems can be evaluated in simulation with the following steps:

1. Generate a random password as the root secret.
2. Generate a challenge for an authentication round.
3. Generate a response based on the password and the system design.
4. Analyze the collected challenge-response pairs after each authentication round assuming that the adversary has the full knowledge of the system design except the password.
5. Repeat steps 2, 3, and 4 until the exact password is recovered.

The security strength of a tested LRP system is measured in this chapter as the average result of 20 runs of simulation.

1.3 Brute Force Attacks and Defense Principles

In the following, we introduce the brute force attacks to LRP systems, and a couple of defense principles to thwart the brute force attacks.

1.3.1 Brute Force Attacks

The brute force attacks can be considered as a general pruning-based learning process, where an adversary keeps removing irrelevant candidates when more and more cues are available. The procedure of the brute force attacks can be described as follows:

1. List all possible candidates for the password in the target LRP system.
2. For each independent observation of a challenge-response round, check the validity of each candidate in the current candidate set by running the verification algorithm used by the server, and remove invalid candidates from the candidate set.
3. Repeat the above step until the size of candidate set reaches a small threshold.

The above procedure shows that the efficiency of the brute force attacks is design-independent. The following two statements reveal the power of the brute force attacks.

Statement 1. *The verification algorithm used in the brute force attacks for candidate verification is at least as efficient as the verification algorithm used by the server for response verification.*

The proof is trivial as the verification process for candidate pruning is essentially the same as the verification process for the server to check for correct responses. It is also possible for an adversary to design a more efficient algorithm if he/she can exploit the correlations, if exist, between candidate passwords.

Statement 2. *The average shrinking rate for the size of a valid candidate set is the same as one minus the average success rate of a random guessing attack.*

The average success rate of a random guessing attack is defined as the probability of generating a correct response by randomly picking a candidate from the candidate set. This is an equivalent definition of one minus the average shrinking rate of the valid candidate set. Given X as the size of the candidate set, and d as the average success rate of a random guessing attack, the average number of rounds to recover the exact secret is $m = \lceil \log_{1/d} X \rceil$, assuming that each candidate is independent of other candidates. If each candidate is not independent, the average number of rounds to recover the exact secret would be smaller than m. This statement can be used to estimate the average success rate of a random guessing attack, $d = X^{-\frac{1}{m}}$. This statement explains why most LRP systems reveal the entire secret after one or two authentication sessions being recorded by an adversary [25], as their expected success rates of a random guessing attack are sufficiently low so that the whole candidate set rapidly collapses to the exact secret. This result implies that lowering the success rate of a random guessing attack is at the cost of leakage resilience against the brute force attacks.

1.3.2   Large Root Secret Space Principle

Principle 1. *An LRP system with non-zero secret leakage should have a large candidate set for the root secret.*

The above defence principle requires a large password space as the basic defense against the brute force attacks, where "large" means that it is computational infeasible for an adversary to enumerate all candidates in a practical setting (the same meaning of "large" is used in the following discussions).

This principle seems trivial, but, it is not always straightforward to determine whether an LRP system results in any secret leakage under a given threat model. In general, there are three possible leakage sources in an LRP system, including *the response alone*, *the challenge-response pair*, and *the challenge alone*. Among them, the last source has not been well recognized. We use Undercover [26] as an example to show that secret leakage could happen even when a secure channel is present.

Undercover is a typical scheme based on the k-out-of-n paradigm. During registration, a user is assigned with k images as his/her secret from a pool of n images. In each authentication round, the user is asked to recognize if there is a secret image from w candidate images and report the position of that image if the secret image is shown in the current window; otherwise, the user reports the position of the "none" symbol. Before the user reports the position, a haptics-based secure channel is deployed to map the real position to a random position decided by the hidden message delivered via the secure channel.

In the design of Undercover, the hidden mapping blinds an adversary from learning any information from a user's input during password entry. It was suggested that a small password space is sufficient with the default parameters $k = 5, n = 28$, and $w = 4+1$ (i.e. four images and a "none" symbol). The number of candidate root secrets is $C_{28}^5 = 98,280$. However, this scheme does not prevent the challenge alone

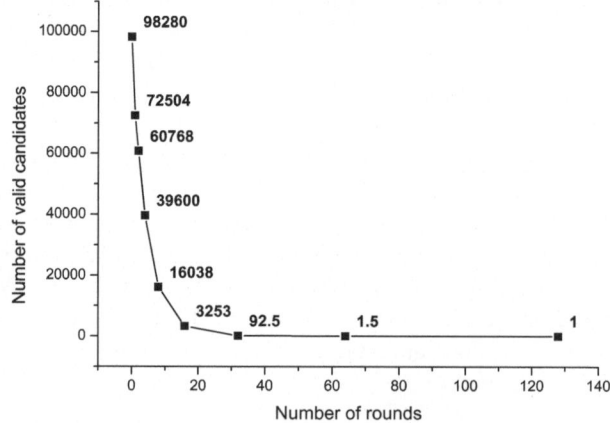

Fig. 1.2 Average number of valid candidates shrinks for Undercover

from becoming a source of leakage. In Undercover, there is at most one secret image taken among w candidate images for each authentication round. This implies that a candidate of the root secret is invalid if at least two images of this candidate appeared in an authentication round. Since the candidate space is small, one can use the brute force attacks to recover the secret according to the challenge alone. Figure 1.2 shows how the size of the candidate space shrinks as the number of observed authentication rounds increases. On average, 53.06 rounds (6 sessions) are sufficient to recover the exact secret, and the size of the candidate set can be reduced to less than 10 after 43.55 rounds (5 sessions). This result shows that a secure channel alone is not sufficient to prevent secret leakage in the design of Undercover. It is necessary to follow the large root secret space principle in this case.

A similar problem applies to the Convex Hull Click (CHC) scheme [31], where the default parameters are $k = 5$, $n = 112$, and $w = 83$. The size of the candidate set for the root secret is $C_{112}^5 = 1.34 \times 10^8$. In simulation experiments, we recover the exact secret within 12.28 rounds (2 sessions). Another interesting finding for CHC is that we can now estimate the average success rate of a random guessing attack from the result of the brute force attacks, though a precise analysis is difficult [31]. According to *Statement 2*, the average success rate is $21.78\% = (C_{112}^5)^{-\frac{1}{12.28}}$.

1.3.3 Large Round Secret Space Principle

Principle 2. *An LRP system with non-zero secret leakage should have a large candidate set for each round secret.*

This large round secret space principle emphasizes that a large candidate set for the root secret is necessary but not sufficient to defend against the brute force attacks. A large candidate set for the root secret can be broken down by effective attacks to the round secrets if the round secrets are different from the root secret. We use the Predicate-based Authentication Services (PAS) scheme [4] as an example to show that a round secret with a small candidate set can be easily recovered and later used to reveal the root secret.

During the registration of PAS, a user is required to remember p secret pairs, each of which comprises a secret position and a secret word. At the beginning of each authentication session, the server shows the user an integer index I. Then the user uses I to calculate p predicates as follows: For each pair, the corresponding predicate is the secret position and a secret character. The secret character is the x-th character in the secret word, where $x = 1+((I-1) \bmod len)$, and len is the length of the secret word. For example, given two secret pairs $(\langle 2,3 \rangle$, sente$)$, $(\langle 4,1 \rangle$, logig$)$ and $I = 15$, the predicates are $(\langle 2,3 \rangle$, e$)$ and $(\langle 4,1 \rangle$, g$)$, where $x = 1 + ((15 - 1) \bmod 5) = 5$, and the secret position $\langle a, b \rangle$ means "at row a and column b". Given these p predicates, the user examines the cells at the secret positions in l challenge tables to check whether a secret character is present in its corresponding cell. It yields an answer vector that consists of $p \cdot l$ "present" or "absent" answers with a candidate space of 2^{pl}. This vector is then used to lookup another response table, which provides a many-to-one mapping from 2^{pl} elements to 2^l elements. Finally, the user inputs one of those 2^l elements indexed by the answer vector to finish an authentication round.

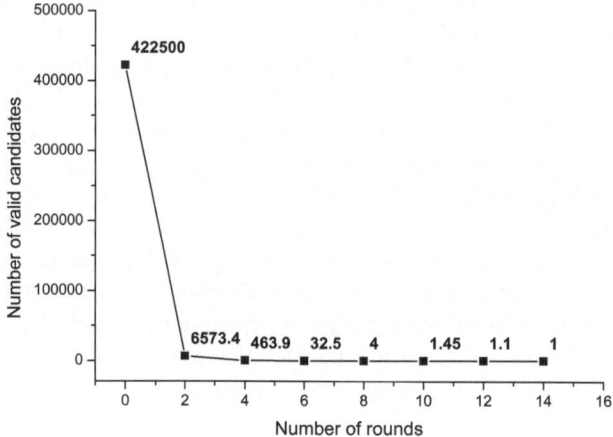

Fig. 1.3 Average number of valid candidates shrinks for PAS

The above many-to-one mapping is used in PAS to confuse the adversary. However, if a round secret is associated with a small candidate set, many mappings would have the same pre-image and the effective mapping space would collapse to the candidate set of the round secret. In PAS, the size of the candidate set for each round secret is $422,500 = (25 \times 26)^2$ in the default case, where $p = 2$, and there are 25 cells in each challenge table and 26 possible letters for the secret character. It is not difficult to use the brute force attacks to recover each round secret of PAS. Figure 1.3 shows the shrinking of the candidate set as the number of observed authentication rounds increases. On average, 9.4 rounds are sufficient to recover the exact round secret (1 session). Since all the predicates generated from the same secret pair share the same secret position, after recovering the first round secret, it is easy for the adversary to recover the other round secrets and finally the root secret. In our simulation experiments, we discover the exact root secret with 8 sessions.

Note that a similar attack has been used in analyzing PAS [18]. This vulnerability also appears in the S3PAS scheme [35], which is a variant of the CHC scheme [31].

1.4 Statistical Attacks and Defense Principles

In the following, we introduce the statistical attacks to LRP systems, and three defense principles to thwart the attacks.

1.4.1 Statistical Attacks

The statistical attacks can be considered as an accumulation-based learning process, where an adversary gradually increases his/her confidence on relevant targets when more and more cues are available. Compared to the brute force attacks, the statistical attacks have fewer limitations as they can be applied to the LRP systems even with a large password space as long as a user's response during password entry is statistically biased towards his/her knowledge of the secret. Theoretically, there exist certain statistical attacks to any LRP system with non-zero secret leakage. The efficiency of the statistical attacks is design-dependent and varies with different schemes and different analysis techniques. We introduce two general statistical analysis techniques which can be used to efficiently extract the root secrets of most existing LRP schemes.

The first technique is named as *probabilistic decision tree*. It works efficiently for the existing LRP systems with simple challenges [4, 30, 31, 35]. The procedure of probabilistic decision tree is given below.

1. Create a score table for each possible individual element or element group in the alphabet of the root secret, where the element group should be computationally feasible to maintain. A *t-element score table* is referred to as a score table whose entry contains t individual elements.
2. For each independent observation of a challenge-response pair, an adversary enumerates all *consistent* decision paths leading to the current response, where the notion of decision path is given in Fig. 1.4. Each possible decision path is assigned with a probability calculated based on the uniform distribution. In the k-out-of-n paradigm, the probability is $p_1 = k/n$ for a decision event in which the corresponding individual element belongs to the secret set, and $p_0 = 1 - p_1$ for the complementary event. For the example decision path X given in Fig. 1.4, its probability is $p(X) = p_1 \cdot (p_0 \cdot p_1)$. After enumerating all consistent decision paths, the adversary sums up the probabilities of these paths and uses the sum p_c to normalize the probability $p(X)$ for each decision path X to its conditional probability $p(X|C) = p(X)/p_c$, where the conditional probability represents the probability that the decision path is chosen by the user when the current response C is observed. After the normalization, the adversary updates the score table using $p(X|C)$: For each entry that appears in a consistent decision path X, its score is added by $p(X|C)$ if the corresponding event is that the entry belongs to the secret set; otherwise its score is deducted by $p(X|C)$.
3. Repeat the above step until the number of entries with certain score levels reaches a threshold (e.g. k 1-element entries with the highest scores).

The second technique is named as *counting-based statistical analysis*. The basic idea of this technique is to simply maintain a counting table for the occurrences of elements. Multiple counting tables may be maintained simultaneously according to different response groups. The procedure proceeds as follows.

Definition: A *decision path* is an emulation of a user's decision process that consists of multiple decision nodes. Each *decision node* represents a decision event decided by whether or not the corresponding entry in the score table belongs to the secret set.

Example: Consider a scheme which shows a four-element window $\langle S_1:1, S_2:2, S_3:1, D_1:1 \rangle$ and requires a user to report the sum of the numbers associated with the *first* and *last* secret elements displayed in the window, where $S_i:x$ represents a secret element associated with number x, and $D_i:y$ represents a decoy element associated with number y. Since the correct response for this challenge is 2, which is the sum of the numbers associated with the first and third elements, its decision path is $X = \langle S_1:1 \rangle | \langle D_1:1; S_3:1 \rangle$. There are two segments in this decision path. The first segment implies that S_1 is a secret element, and the second segment implies that D_1 is a decoy element and S_3 is a secret element. There usually exist other decision paths leading to the same response, such as $\langle S_1:1 \rangle | \langle D_1:1 \rangle$.

Fig. 1.4 Definition and example of decision path

1. Create l counting tables for l response groups. An adversary may create a counting table for each possible response if it is affordable. "Any response" is a useful response group if the secret elements appear more or less frequently than the decoy elements *in the challenge*. An entry in a counting table can be any individual element or element group. A *t-element counting table* is referred to as a counting table whose entry contains t individual elements. If $t \geq 2$, this statistical analysis is named as *multi-dimensional counting*.

2. For each independent observation of a challenge-response pair, the adversary first decides which counting table is updated according to the observed response. Then each entry in the chosen counting table is incremented by the number of occurrences of the corresponding individual element or element group. If the group of "any response" is used, its counting table is always updated for each observation.

3. Repeat the above step until the number of entries with different score levels reaches a threshold. The score for an entry is a weighted sum of the count values for the same entry in different tables. The weight function is dependent on the LRP system design and the response grouping strategy.

1.4.2 Uniform Distributed Challenge Principle

Principle 3. *An LRP system with non-zero secret leakage should make the distribution of the elements in each challenge as uniform as possible.*

The above uniform distributed challenge principle requires that an LRP system should generate the challenges without knowing the secret. The secret, or any value derived from the secret, should be only used by the server to verify the responses.

If there is any *structural requirement* in the challenge generation, then it is likely that secret leakage would happen. Non-uniformly distributed elements in a challenge may leave cues for an adversary to recover the secret even without knowing the response. In the following, we examine how Undercover [26] leaks secrets from biased challenges.

Undercover ensures that the distribution for each image is unbiased by showing every candidate image exactly once for each authentication session. However, its two-dimensional distribution is biased in each authentication round, as no secret-secret pairs appear in any challenge (at most one secret image appearing). An adversary may use a *2-element counting table* to recover the secret from the challenges. For each pair of candidate images, the count value is zero only if both of them belong to the secret set after a sufficient number of observations. On average, it is sufficient to recover the exact secret within 172.7 rounds (or 20 sessions), and recover 80 % of secret elements (five secret images in total) after 126.9 rounds (or 15 sessions).

This attack applies to the CHC scheme [31] and the low-complexity CAS scheme [30] as well. Both of them require that at least k secret elements appear in the challenge window, while the challenge window holds only a subset of candidate elements. Such structural requirements make the distribution of the elements in each challenge deviates from the uniform distribution. With default parameters, an adversary may recover the exact root secret within 18.18 rounds (or 2 sessions) for CHC. For the low-complexity CAS scheme, an adversary may recover the exact root secret (i.e. 60 independent secret images) within 2,087.2 rounds (or 105 sessions), and recover 90 % of secret elements within 870.4 rounds (or 44 sessions).

According to the uniform distributed challenge principle, it is better to make the distribution of the elements in each challenge indistinguishable from the uniform distribution so as to prevent the password leakage from biased challenges. One simple (but not necessarily good) way to achieve this is to display all elements in each challenge.

1.4.3 Indistinguishable Individual Principle

Principle 4. *An LRP system with non-zero secret leakage should make each individual element indistinguishable in the probabilistic decision tree if the candidate set of decision paths is enumerable.*

The above indistinguishable individual principle is critical to restrict the probabilistic decision tree attack. The power of probabilistic decision tree stems from its emulation of all possible decision processes leading to the observed response. The emulation creates a tight binding between each challenge and its response if individual elements are distinguishable on consistent decision paths. It is challenging to make each individual element indistinguishable, especially when certain weight or order information is used in the challenge design. In the following, we examine the high-complexity CAS scheme [30] and show how probabilistic decision tree can be used to discover the root secret efficiently even when a number of decision paths lead to the same answer.

The high-complexity CAS scheme is another typical LRP scheme in the k-out-of-n paradigm. During registration, a user is assigned with $k = 30$ images as his/her secret from a pool of $n = 80$ images. In each authentication round, a challenge is a 8×10 grid consists of all 80 images, one image for each cell. The user is asked to mentally compute a path starting from the cell in the upper-left corner. The computation rule is described as follows:

- Initially the current cell is the cell in the upper-left corner.
- If the image in the current cell belongs to the secret set, move down by one cell; otherwise move right by one cell.
- If the next moving position is out of the grid, it is referred to as an *exit position*.
- The path computation ends with an exit position.

The user reports the answer associated with that exit position to finish an authentication round, where the answer is an integer from $[0, 3]$, and is randomly assigned to each exit position. Since the same answer is assigned to multiple exit positions (i.e. 4 answers assigned to 18 exit positions), an adversary cannot easily tell which the exact exit position is. For each exit position, there are many possible paths leading to it, which makes it difficult for the adversary to identify the exact path.

Since the default parameters are large enough ($k = 30$, $n = 80$), a brute force attack to this scheme is not feasible in the current stage. The scheme follows *Principle 3* in displaying all the candidate images in each challenge; therefore, an adversary cannot extract the secret by analyzing the challenges only. Nonetheless, each individual element is not indistinguishable in this scheme during the decision process, as different elements may have different impacts on the transition of decision paths. An adversary can thus use the probabilistic decision tree to recover the secret from the observations of some challenge-response pairs.

Each possible path leading to an observed response forms a decision path in the probabilistic decision tree. The probability of a decision path is decided by the movements on this path. For example, a path $X = \langle DOWN, RIGHT, RIGHT, DOWN \rangle$ means the first and the fourth images belong to the secret set, while the second and third images do not. The probability $p(X)$ is $p_1 \cdot p_0 \cdot p_0 \cdot p_1$, where $p_1 = k/n$ and $p_0 = 1 - p_1$. Initially, we create a *1-element score table*. Given a response with the answer i, one may enumerate all consistent decision paths leading to this answer, and update the score table according to the conditional probability $p(X | \text{response} = i)$.

For an 8×10 grid specified under the default parameters, there are 43,758 possible decision paths in total, with the average path length 14.5539. For each candidate image, its score is at a significantly high level if it belongs to the secret set after a sufficient number of observations. Figure 1.5 shows the false positive rate decreasing as the number of observed authentication rounds increases. On average, it is sufficient to discover the exact secret within 640.8 rounds (or 65 sessions), and discover 90 % secret elements after 264.7 rounds (or 27 sessions). Although the

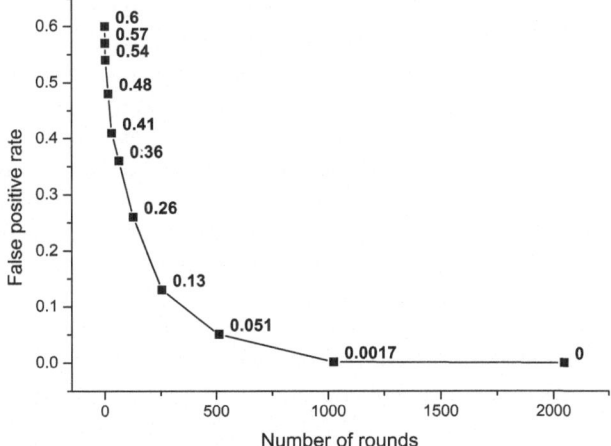

Fig. 1.5 Average false positive rate of high-complexity CAS scheme

required number of observed sessions is not small, it is still possible for an adversary to collect them using a key logger, and such security strength is achieved only if a user can remember 30 independent secret images.

The probabilistic decision tree attack also applies to the low-complexity CAS scheme [30], the CHC scheme [31], the S3PAS scheme [35], and the PAS scheme [4]. All of these schemes have an enumerable candidate space for decision paths, and different elements may have different impacts on the transition of decision paths.

These examples show that it is necessary to increase the number of candidate decision paths if it is infeasible to make each individual element indistinguishable in the probabilistic decision tree. The only known designs that satisfy this indistinguishability requirement are the counting-based schemes [15, 20]. In those schemes, there is no order or weight information associated with each candidate element, which usually distinguishes the elements in decision paths. The user is asked to count the secret elements appearing in a challenge. The response is generated based on the count value. In such case, the probabilistic decision tree attack does not apply; nonetheless, it is still possible that they are subject to the counting-based statistical analysis attack (see Sect. 1.4.1).

1.4.4 Indistinguishable Correlation Principle

Principle 5. *An LRP system with non-zero secret leakage should minimize the statistical difference in low-dimensional correlations among possible responses.*

The indistinguishable correlation principle is complementary to *Principle 4* in thwarting the counting-based statistical analysis. Although the counting-based statistical analysis is straightforward, it cannot be completely prevented if the user's response is always statistically biased towards his/her knowledge of the secret in password entry. In an extreme case, an adversary may maintain a counting table for all candidates of the root secret, and update the table according to available observations. With the help of such counting table, the adversary may identify the statistical difference in responses leading to the root secret even when the user is asked to make intentional mistakes at a predefined probability known by the server only. An explanation is given below.

Consider the following assumption: (i) The user makes mistakes in the responses with a fixed error probability ρ; (ii) the average success rate of guessing attack on the "correct" response for each authentication round is d; and (iii) the number of candidate root secrets is N. Under such assumption, an adversary cannot distinguish the true secret only if the equation $\frac{1-\rho}{(1-\rho)(Nd-1)+\rho \cdot Nd} = \frac{1}{N-1}$ holds, meaning that the decoys get the same count value as that of the secret. Solving this equation yields $\rho = 1 - d$, meaning that the user should provide the correct response with probability $1 - \rho = d$. This implies that the user's decision process is the same as a random guessing, which defeats the purpose of authentication.

The counting-based statistical analysis is effective if sufficient resources are available to an adversary. In reality, the resources available to the adversary are not unbounded. The cost of maintaining a *t-element counting table* is $O(n^t)$, which increases exponentially with the number of elements t contained in a table entry, where n is the number of total individual elements. If the adversary fails to maintain a high-dimensional counting table, the correlation information in this table may not be obtained by the adversary. However, it is still possible for the adversary to exploit the low-dimensional correlation to recover the secret. We use SecHCI [20] as an example to show how it works while the brute force attack and the probabilistic decision tree attack are infeasible.

During registration of SecHCI, a user is assigned with k icons as his/her secret from a pool of n icons. In each authentication round, the challenge is a window consisting of w icons. The user is asked to count how many secret icons appearing in the window. After getting the count value x, the user calculates $r = \lfloor (x \bmod 4)/2 \rfloor$. The final response r is either 0 or 1. The challenge is designed so that each individual candidate has the same probability to appear in the window for either response 0 or 1. Hence, it is impossible for an adversary to extract any useful information from the 1-element statistical analysis.

Since the default parameters $k = 14$ and $n = 140$ are sufficiently large, it is infeasible for an adversary to use the brute force attack to break SecHCI. Also because it is a counting-based scheme, it is not subject to the probabilistic decision tree attack according to *Principle 4*. However, a two-dimensional counting attack is still applicable. Compared to decoy icons, there are 0.599 more pairs on average among the secret icons for response 0, and 0.599 less pairs on average among the secret icons for response 1. An adversary may use two *2-element counting tables* to recover the secret, one table for each response. The adversary may update the count value for each pair displayed in each challenge and each response. The score for each entry is calculated as the value difference between these two tables. For each pair of candidate icons, the score is at a significantly high level if both of them belong to the secret set after a sufficient number of observations.

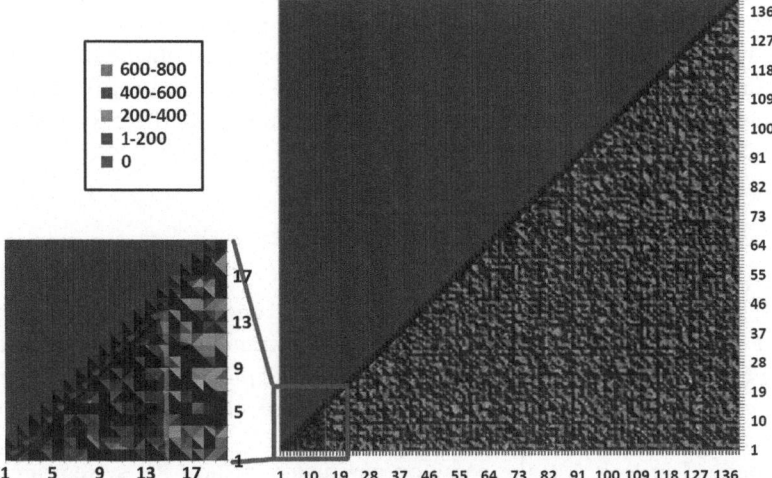

Fig. 1.6 Pair-based score distribution for SecHCI (the first 14 elements are the secret icons, whose pair-based scores are distinguishable from the scores of other icons)

Figure 1.6 shows the pair-based score distribution for SecHCI after 20,000 authentication rounds, from which the secret icons can be easily distinguished. On average, it is sufficient to recover the exact secret with 14,219.4 rounds (or 711 sessions), and recover 90 % secret elements after 10,799.8 rounds (or 540 sessions). Since SecHCI follows most of our principles, these numbers are much larger than the schemes we analyzed previously, but it is still far less secure than it was claimed [20]. On the other hand, the security strength of SecHCI is achieved by imposing a high cognitive workload to a user where the user is asked to correctly examine 600 icons (30 icons per round × 20 rounds) one by one for each authentication session.

The secret leakage of SecHCI due to pair-based statistics can be fixed by changing its response function from $r = \lfloor (x \bmod 4)/2 \rfloor$ to $r = x \bmod 2$, where x is the number of secret icons in each challenge window; however, this fix would make SecHCI subject to the algebraic attack based on Gaussian elimination [20]. This is also the original motivation of SecHCI using the current function. To thwart this algebraic attack, a user may produce incorrect answers with a fixed error probability as suggested in [15]; however, this would decreases the scheme's usability.

Another design limitation on counting-based LRP schemes, including SecHCI, is that the response function cannot be designed in the form of $r = x \bmod q$ if q is an integer larger than 2. The reason is that in our simulation experiments, we discover that the pair-based statistical difference appears significant when q is larger than 2, and it increases with the value of $|r - w \cdot k/n|$, where r is the response value, w is the window size, k is the number of secret elements, and n is the total number of elements. This can be explained as follows. For a response, if the expected number of secret elements in a window is less than the expected number $w \cdot k/n$ derived from the uniform distribution, then the number of pairs among secret elements is also less than the expected number $C^2_{wk/n}$, and the number of pairs among decoy elements is larger than the expected number derived from the uniform distribution, and vice versa. In such case, an adversary can distinguish the secret elements from the other elements by grouping them according to the observations of different responses.

1.5　Usability Costs of Defense Principles

In this section, a qualitative analysis is provided on the usability costs of the defense principles given in Sects. 1.3 and 1.4, showing the relations and tradeoffs among the constraints imposed by the defence principles and the requirements on human capabilities. A quantitative analysis framework on the usability costs of LRP schemes will be provided in the next section.

As defined in Sect. 1.2, the common parameters of an LRP system are given in a tuple (D, k, n, d, w, s). All of the parameters except D (i.e., the expected authentication strength) are affected by the defence principles.

Principles 1 and 2 require a large candidate set for the root secret and the round secret. This implies that either k (i.e., the number of secret elements) or n (i.e., the total number of candidate elements) should be sufficiently large. An increase in k requires the user to memorize more elements as his/her secret. While an increase in n does not raise the user's memory demand, it increases the statistical significance of the secret in the whole candidate set, and thus indirectly increases the user's computation workload as analyzed later.

Principle 2 also raises the user's computation workload, as it indicates that a challenge is vulnerable to the brute force attacks if it can be solved using a small number of possible secret elements. In order to increase the candidate space of the round secret, the round secret must be either randomly selected from the root secret [20, 30, 31] or use all elements in the root secret [2, 15]. The former choice requires the user to recognize the current displayed secret elements which change randomly in every round, and the latter requires the user to recall a large number of secret elements. Also, more elements appearing in a challenge means more computation workload to aggregate them into the correct response.

Principles 3, 4, and 5 have more impact on (d, w, s), where d is the average success rate of guessing attack in a single round, w is the average window size for each authentication round, and s is the average length of user's decision path. Principle 3 requires that the elements in a challenge should be uniformly drawn from the candidate set. Due to previous requirements of large secret space and our preference of minimizing the demand on user's memory, the value of k should be small and the value of n should be large. The consequence is that the average number of secret elements displayed in a challenge window, $w \cdot k / n$, cannot be too large if the window size w is not large enough. This restricts the number of possible responses to a small value, which raises the success rate d of a guessing attack and increases the round number for an expected authentication strength D. On the other hand, if the window size is large, an LRP system can only be implemented on large-screen devices and it also increases the difficulty of user's examining the elements in a challenge window. Regardless of the window size, this principle may impose a high computation workload and a high error rate for users.

Principles 4 and 5 further rule out most existing LRP schemes which use simple challenges. Principle 4 states that if the design of a challenge is not complex enough to aggregate a large number of secret elements into a response, it leads to a counting problem. Principle 5 further states that only 0 and 1 can be safely used as a response so as to address the counting problem if the modular operation is the only operation used for generating the final response. Consequently, there are three possible choices which are left for forming a challenge:

1. A complex challenge with many secret elements. In this case, the round number would be small but the challenge is rather difficult for a user to respond; in other words, the average length s of decision paths would be long.
2. A counting-based challenge with its response generated using the modular operation only. In this case, the round number would be large and the challenge is relatively easier to respond.
3. A counting-based challenge with its response generated using a specially designed function which has a large number of possible responses and satisfies the correlation indistinguishability condition. In this case, the challenge is to design such function with acceptable usability.

It is clear that all of the choices impose a considerable burden on users.

1.6   Tradeoff Between Usability and Security

In this section, we first establish a quantitative analysis framework for evaluating the usability cost of typical LRP systems. This framework decomposes the process of human-computer authentication into atomic cognitive operations in psychology. There are four types of atomic cognitive operations commonly used: single/parallel recognition, free/cued recall, single-target/multi-target visual search, and simple cognitive arithmetic. The definitions and performance models of these atomic cognitive operations are given in the Appendix of this chapter.These performance models are used to evaluate the cognitive workload of average human users of LRP systems. Based on the usability costs, a quantitative assessment of the tradeoff between usability and security is provided for analyzing LRP systems.

According to the conventions in the psychology literature, "user" is also referred to as "subject" in this section.

1.6.1   Quantitative Analysis Framework of Usability Costs

There are two components in our quantitative analysis framework for the usability costs of LRP systems: *Cognitive Workload* (C) and *Memory Demand* (M). The cognitive workload is measured by the total reaction time required to perform the cognitive operations in LRP schemes. A long reaction time for each authentication round means that it is difficult for a subject to answer each challenge; it also means that the overall error rate would be high. A long reaction time for each authentication session means that the overall cognitive workload is high and that the involvement of attention and patience is also high. On the other hand, the memory demand is measured by the number of elements which should be memorized by a subject for authentication. Since this memorization process is independent from the authentication process, we consider it as a separate component.

For cognitive workload, the cost of an authentication round is the sum of average reaction time for performing all atomic cognitive operations in the round. This cost represents the average thinking time of a subject required to answer a challenge. A typical authentication round consists of at least a memory retrieval operation and a simple arithmetic operation. For graphic-based schemes, a visual search operation is also common. According to the working memory capability theory [9, 24, 28, 29], the average reaction time cannot be shortened by repetitive rehearsals if a subject maintains more than $4(\pm 1)$ items in his/her working memory. A rehearsal may only improve the accuracy of a subject's response, which represents an inherent limitation of human capabilities. This limitation also applies to other non-memory operations such as visual search where the item positions are shuffled in each challenge [32]. Overall, the cognitive workload of an authentication session can be calculated as the product of the cognitive workload of an authentication round and the round number when the number of the secret items is larger than 5. For the schemes [4, 31] with no more than 5 secret items, we only count once for their memory retrieval operations, assuming that the secrets are not flushed out according to the working memory capacity theory.

Besides the reaction time, other cognitive workload measurements such as user frustration level, concentration load, and motivational effort are usually collected from standardized testing questionnaires. However, these measurements are susceptible to various implementation and environmental factors, such as screen size, graphic or text-based interface design, and the education background of subjects. In contrast, the influence of those unstable factors has been minimized in more than a century's development of experimental psychology. A major advantage of using the performance models of atomic cognitive operations in our analysis is that the models are *implementation-independent*, which is necessary for a fair comparison between different LRP designs. Our estimation of the cognitive workload of typical existing LRP schemes is consistent with, or very close to, the time costs reported in the original papers [4, 20, 30, 31].

For memory demand, the cost of an LRP scheme is the ratio k/λ_{op} between the number of secret items, k, and the *accuracy rate* of corresponding memory retrieval operation within a fixed memorization time, λ_{op}. Since recognition is much easier than recall [10, 14, 23, 24, 28], it is necessary to distinguish different memory retrieval operations. According to [14], λ_{op} is 29.6 % for recall and 84.8 % for recognition.

One may argue that a better estimation for the memory demand would be the minimum time for a subject to remember all secrets for using an LRP system. However, the lower bound of such memorization time is difficult to measure in experimental psychology, as the subject may not realize the precise time for him/her to remember all of the required secrets. A less confident subject may take more time than what is actually required. For a similar reason, no other memory factors, such as password interference and recall accuracy over extended periods, are integrated in our current analysis framework.

Finally, an overall score, HP (standing for *Human Power*), is calculated as the product of cognitive workload score HP(C) and memory demand score HP(M). This score (HP) indicates the expected human capability requirement for a human-computer authentication system.

1.6.2 High Security at Cost of Heavy Cognitive Demand

Table 1.1 shows the cognitive workload and the security strength of typical LRP systems. The LRP systems are listed in the descending order of their HP scores. The parameters in these LRP schemes are set to their default values except that the round number is adjusted so as to make the successful rate of a random guessing attack at the same level as the authentication strength of a 6-digit PIN. This adjustment is necessary to make a fair comparison of the schemes based on the same strength of defending against an adversary without any prior knowledge. The other two points in this table which need explanations are about PAS [4] and CHC [31]. In PAS, we consider the root secret for each authentication session as predicates instead of complete secret pairs, since the same predicates are used for all rounds in an authentication session. In CHC, the expected successful rate of a random guessing attack is not reported in the original paper. We estimate it based on *Statement 2* (see Sect. 1.3.1), and derive it to be 21.78 % from our simulation results.

The column "HP(C)/round" in this table shows the cognitive workload required to solve a challenge in each authentication round, where the calculation formulae of this cognitive workload for different LRP schemes are given in Table 1.2 in the Appendix of this chapter. All of the results except for LPN [15] and APW [2] are very close to the average time costs reported in the original papers [4, 20, 30, 31]. For LPN, there was no report on a controlled user study. In the original paper, LPN was implemented as a public web page, to which subjects could freely access and get rewards for each successful login. There was no evidence showing that the subjects were asked to memorize their root secrets (i.e., 15 secret positions), and then recall them in each authentication round. Thus, the average time cost

Table 1.1 Usability and security tradeoff of typical LRP systems with default parameters

	k	n	Win size	Password space	Guess rate/round	No. of rounds/login	Reported time/round (s)	HP (C)/rounds(s)	HP (C)/login(s)	HP (M)	HP = M × C ($\times 10^2$)
LPN [15]	15	200	200	1.463×10^{22}	0.50	20	23.71	33.423	668.45	50.68	**338.74**
APW [2]	16	200	200	8.369×10^{24}	0.10	6	35.50	57.928	347.57	54.05	**187.87**
CAS low [30]	60	240	20	2.433×10^{57}	0.50	20	5.00	6.073	121.46	70.75	**85.94**
CAS high [30]	30	80	80	8.871×10^{21}	0.25	10	20.00	22.099	220.99	35.38	**78.18**
SecHCI [20]	14	140	30	6.510×10^{18}	0.50	20	9.00	10.638	212.76	16.51	**35.13**
CHC [31]	5	112	83	1.341×10^{8}	0.22	10	10.97	9.326	93.26	16.89	**15.75**
PAS [4]	4	N/A	13	4.225×10^{5}	0.25	10	8.37	6.837	68.37	13.51	**9.24**

reported for each round was likely underestimated. In the original paper of APW, the cognitive workload was directly estimated based on the results of LPN (with no actual user study conducted); thus, it could also be underestimated.

Table 1.1 shows three tiers of LRP schemes. From bottom to top, the schemes in an upper tier have better security strength against secret leakage at the cost of lower usability. The schemes in the bottom tier are PAS [4] and CHC [31], which are susceptible to both brute force and statistical attacks. Moving to the middle tier of CAS [30] and SecHCI [20], the memory demands are higher, and the brute force attacks become infeasible. However, CAS and SecHCI are still susceptible to the statistical attacks. The main reason is that the simple challenges used in these schemes can still be used to reveal the statistical significance of secret passwords. More cognitive workload is thus required to mix secret items with other items in the password entry process. The top tier consists of LPN [15] and APW [2], which follow all of our defence principles. They are immune to both brute force attacks and statistical attacks in practical settings; on the other hand, they impose significantly high usability costs.

There is an interesting finding when we look at the two schemes in the top tier. In the quantitative analysis framework, LPN has a higher HP score but a smaller password space compared to APW. This is because the security measurement is limited to the brute force attacks and statistical attacks. It might be possible in the future that other more efficient attacks are discovered and thus lower the security strength of APW. The tradeoff between usability and security in the quantitative analysis framework may not strictly follow the order of HP, as it is always feasible to design an LRP scheme with a lower usability for a given security strength. The meaning of the quantitative analysis framework is that the human capability of subject should reach a *lower bound* so as to achieve a high security strength.

Our analysis reveals an inherent limitation in the design of LRP systems. The high cognitive workloads indicate the incompetence of human cognitive capabilities in using secure LRP systems without leveraging any secure channels in practical settings. This may explain why the problem of designing highly secure and usable LRP systems is still challenging since its first proposal [22] 20 years ago.

1.7 Bibliographic Note

Since passwords are pervasively used in the modern society, the design of secure and usable password systems has been extensively investigated. We summarize closely related works in several aspects, including attacks, principles, and the tradeoff analysis for LRP systems.

Most of existing LRP systems are broken. The recent works on the attacks to LRP systems include the following: (i) Golle and Wagner proposed the SAT attack [13] against the CAS scheme [30]; (ii) Li et al. demonstrated the brute-force attacks [18] against the PAS scheme [4]; (iii) Li et al. later presented a Gaussian elimination-based algebraic attack [19] against the virtual password system [17];

(iv) Asghar et al. introduced a statistical attack [1] against the CHC scheme [31]; and (v) Dunphy et al. analyzed a replay-based shoulder surfing attack against the recognition-based graphical password systems under a weak threat model [11]. Compared to these works, we present two types of generic attacks, including the brute-force attacks and the statistical attacks, to analyze some existing LRP systems such as Undercover [26] and SecHCI [20], and discover more vulnerabilities as reported in [34].

Some principles have been proposed for designing LRP systems. Roth et al. [25] proposed a basic principle of using a cognitive trapdoor game, which requires that the knowledge of a secret should not be directly revealed during the password entry. Li and Shum [20] later suggested another three principles regarding time-variant responses, randomness in challenges and responses, and indistinguishability against the statistical analysis. Extending the coverage of the previous principles, we report five more principles for thwarting the brute force attacks and statistical attacks in the design of LRP systems [34].

It is challenging to evaluate the usability of LRP systems quantitatively. As pointed out by Biddle et al. [5], the usability evaluation of many LRP systems lacks consistency, which makes it difficult to compare different results. The quantitative analysis framework presented in this book was the first attempt to provide a uniform usability measurement based on experimental psychology [34]. Based on this framework and security analysis, a strong tradeoff between security and usability is discovered, which indicates an inherent limitation in the design of LRP systems. This limitation was first addressed by Hopper and Blum [15], anticipating that practical solutions for unaided humans meeting both security and usability requirements could be discovered in the future. Unfortunately, it was revealed that such solutions do not exist if no secure channel is exploited during password entry [34]. Coskun and Herley [8] also reached a similar conclusion by analyzing the efficiency of brute force attacks with regards to the entropy of user responses during password entry. Their conclusion is based on the assumption that a user must make a large number of sequential binary decisions so as to generate a high response entropy. However, this assumption may not be always valid as humans have a strong capability of processing certain visual tasks (e.g. visual search) in parallel.

We would like to point out two limitations of the current usability framework. First, since the cognitive workload may not be independent to the memory demand, it is possible to improve the overall calculation of usability instead of using the product of cognitive workload and memory demand (i.e. $HP = C \times M$). Second, the factor of error rate is currently not included in the usability framework since it is not clear in experimental psychology what the general relation is between a user's response time and his/her error rate. The factor of error rate may be investigated in the future so as to improve the precision of this framework.

1.8 Conclusion

In this chapter, we analyzed the inherent tradeoff between security and usability in designing an LRP system without using any secure channel. First, we summarized the impacts of two generic attacks, including brute force attacks and statistical attacks, on typical LRP schemes designed for unaided humans. Unlike any specific attacks such as SAT [13] and Gaussian elimination [19], these two types of attacks are generic, which cannot be mitigated with low human capabilities. Then, we introduced five principles necessary to achieve leakage resilience in password entry. The usability costs for these principles indicate that a secure LRP system without a secure channel incurs either high memory demand or high cognitive workload for unaided humans. Finally, we introduced a quantitative analysis framework to further understand the tradeoff between security and usability, showing that an average unaided human may not be competent enough to use a highly secure LRP system without leveraging any secure channel in practical settings.

Appendix: Atomic Cognitive Operations and Human Cognitive Workload Calculation

Table 1.2 Cognitive workload for typical LRP systems ($\alpha_0 = 0.738$, $\alpha_1 = 0.773$, $\alpha_2 = 0.959$, $\alpha_3 = 0.924$ are the average reaction time for arithmetic problems involving 0 or 1, small addition, small division, and large addition, respectively; $\varphi = 1.969$ is the ratio of cued recall compared to single item recognition; $\gamma = 1.317$ is the additional penalty caused by simultaneously recalling the position of an item. For CAS Low and High, 7.4038 and 14.5539 are the average lengths of their decision paths, respectively)

	Atomic cognitive operations	Cognitive workload (HP (C)) per round
LPN [15]	Cued-recall with position, counting, mod	$(0.3964 + 0.0383 \cdot k \cdot \varphi \cdot \gamma) \cdot k + (k/2 - 1) \cdot \alpha_0 + 1 \cdot \alpha_0$
APW [2]	Cued-recall with position, large addition, mod	$((0.3694 + 0.0383 \cdot k \cdot \varphi \cdot \gamma) + 1 \cdot \alpha_3 + 1 \cdot \alpha_0) \cdot k$
CAS low [30]	Parallel recognition, xor	$(0.3694 + 0.0383 \cdot k) \cdot \lceil 7.4038/4 \rceil + 1 \cdot \alpha_0$
CAS high [30]	Recognition	$(0.3694 + 0.0383 \cdot k) \cdot 14.5539$
SecHCI [20]	Parallel Recognition, counting, mod, small division	$(0.3694 + 0.0383 \cdot k) \cdot (\lceil 30/4 \rceil) + 2 \cdot \alpha_0 + 1 \cdot \alpha_0 + 1 \cdot \alpha_2$
CHC[31]	Cued-recall, Multi-target visual-search (3-based)	$((0.3694 + 0.0383 \cdot k \cdot \varphi) \cdot 5/10) + (0.583 + 0.0529 \cdot 83) \cdot 1.8$
PAS [4]	Cued-recall, single-target visual-search, small addition	$(0.3694 + 0.0383 \cdot 2 \cdot \varphi) \cdot 4/10 + (0.583 + 0.0529 \cdot 13) \cdot 4 + 2 \cdot \alpha_1$

There are four types of atomic cognitive operations commonly used in human-computer authentication systems. Their definitions and performance models are introduced in this appendix, which characterize the relation between experiment parameters and reaction time (RT) of an average human. These performance models can be used to evaluate the cognitive workload for typical LRP systems, as shown in Table 1.2.

Single/Parallel Recognition

Recognition is the process to correctly judge whether a presented item has been encountered before. Recognition can be considered as a matching process of comparing presented items with those stored in memory. The reaction time of a recognition operation depends on the number of items which a subject memorizes. The item set in the subject's memory is referred to as a *positive set*. For single item recognition, that is, only one item is shown to the subject each time, one of the most well-known recognition models [27] evaluates the reaction time as $RT = 0.3964 + 0.0383 \cdot k$, where k is the size of the positive set. When multiple items are present simultaneously, the subject can perform recognition in parallel. According to the working memory capacity theory [9, 12, 29], the maximum number of parallel recognition channels is limited to 4 for an average subject. The reaction time of recognizing x items displayed simultaneously can be estimated as $RT = (0.3964 + 0.0383 \cdot k) \cdot \lceil x/4 \rceil$.

Recognition is a common operation in LRP designs. Recognition is used by a subject to judge whether an element appearing in the challenge belongs to the positive set. The high-complexity CAS scheme [30] is an example for single item recognition, where the subject is asked to recognize an image in the current position before deciding which image should be recognized in the next move. The low-complexity CAS scheme [30] and SecHCI [20] are examples of parallel recognition. In the low-complexity CAS scheme, the subject is required to find out the first and the last secret images appearing in a window consisting of 20 images, while in SecHCI, the subject is required to identify all of his/her secret images among 30 candidate images.

Free/Cued Recall

Besides recognition, recall is the other principal method of memory retrieval [3], which is defined as reproducing the stimulus items. Compared to recognition, the recall process is much slower [10, 23]. The common interpretation is that recall is associated with greater resource costs than recognition [10]. Recall might be carried out as a slow process of serial search while recognition as a fast process of parallel retrieval [23].

Free recall and cued recall are two basic recall types. In free recall, a subject is given a list of items to remember and then is tested by recalling them in any order [24]. In cued recall, the subject is given a list of items with cues to remember, and the cues are given in the test. The cues act as guides to what the subject is supposed to remember. For example, given "a body of water", the phrase is the cue for the word "pond" [10]. Many psychological experiments have shown that the reaction time of free recall increases *exponentially* as the size of positive set [24, 28]. In contrast, the reaction time for cued recall is much shorter and it increases *linearly* [10, 23].

Some LRP systems require a subject to recall all his/her secret items during an authentication process. The LPN scheme [15] and the APW scheme [2] are two examples, where the subject is required to recall all the secret items and their corresponding locations in order to identify a challenge digit associated with each secret item. These recall processes should be classified as free recall since no cues are presented. However, no experimental data have been provided in the psychology literature for a large positive set of 15 items as required by these schemes, while the common size for a positive set is 8 for free recall. Since it is difficult to decide whether the exponential trend still holds when the positive set is large, we use the reaction time of the cued recall as a conservative estimate for the free recall which is used in these schemes. According to the experimental results in [7, 23], the formula for the reaction time of the cued recall is $RT = (0.3964 + 0.0383 \cdot \varphi \cdot \gamma \cdot k)$, where φ is the ratio of the cued recall compared to single item recognition ($\varphi = 1.969$ in [23]), while γ is the additional penalty if the subject is required to simultaneously recall the position of an item ($\gamma = 1.317$ in [7]).

Single-Target/Multi-Target Visual Search

Visual search is a perceptual task that involves an active scan of the visual environment for particular targets among other distracters. The measure of the attention in visual search is often manifested as a slope of the function of response time over the number of items displayed (which is referred to as *window size*) [32]. For a single-target visual search, which is the search of a single target among a set of items, its reaction time is believed to be linear to the window size [32, 33] and it can be estimated as $RT = 0.583 + 0.0529 \cdot w$ [33], where w is the window size. For a multi-target visual search, the reaction time is accelerated as the number of targets increases in a fixed-sized window [16].

Visual search is usually used in LRP systems which use simple challenges. PAS [4] and CHC [31] are two examples of using single-target visual search and multi-target visual search, respectively. In PAS, a subject is asked to scan a table cell containing 13 random letters and check whether a secret letter is present or not. In CHC, a subject is required to locate 3 secret elements in a window so as to form a triangle. According to the results given in [16], the reaction time of 3-targets visual search in CHC is approximately 1.8 times longer than that of single-target visual search with the same window size.

Simple Cognitive Arithmetic

Simple cognitive arithmetic is a mental task to solve simple problems involving basic arithmetic operations (e.g., $3 + 4$, $7 - 3$, 3×4, $12 \div 3$). The simple arithmetic problems can be further divided into three subsets: small, large, and zero-and-one problems [6]. For both addition and multiplication, small problems are defined as those with the product of two operands smaller than or equal to 25, and large problems are defined as those with the product of two operands larger than 25. The small and large problems in subtraction and division are defined on the basis of the inverse relationships between addition and subtraction and between multiplication and division. Zero-and-one problem is defined as involving 0 or 1 as an operand or answer. The common instances of zero-and-one problems include counting, exclusive-or, and mod 2. As reported in the experiments of [6], the average reaction time is 0.773 s for small addition, 0.959 s for small division, 0.924 s for large addition, and 0.738 s for zero-and-one problems.

Simple cognitive arithmetic is usually used in the counting-based schemes [15, 20], where a subject is asked to count the number of secret icons appearing in a challenge, and use the count value to calculate a response based on a simple algebraic function.

References

1. Asghar, H.J., Li, S., Pieprzyk, J., Wang, H.: Cryptanalysis of the convex hull click human identification protocol. In: Proceedings of the 13th international conference on information security, pp. 24–30 (2010)
2. Asghar, H.J., Pieprzyk, J., Wang, H.: A new human identification protocol and coppersmith's baby-step giant-step algorithm. In: Proceedings of the 8th international conference on applied cryptography and network security, pp. 349–366 (2010)
3. Baddeley, A.D.: The Essential Handbook of Memory Disorders for Clinicians, Chapter 1, pp. 1–13. Wiley, New York (2004)
4. Bai, X., Gu, W., Chellappan, S., Wang, X., Xuan, D., Ma, B.: Pas: predicate-based authentication services against powerful passive adversaries. In: Proceedings of the 2008 annual computer security applications conference, pp. 433–442 (2008)
5. Biddle, R., Chiasson, S., van Oorschot, P.C.: Graphical passwords: learning from the first twelve years. In: Technical Report TR-11-01 (2011)
6. Campbell, J.I.D., Xue, Q.: Cognitive arithmetic across cultures. J. Exp. Psychol. Gen. **130**(2), 299–315 (2001)
7. Corbina, L., Marquer, J.: Effect of a simple experimental control: the recall constraint in Sternberg's memory scanning task. Eur. J. Cogn. Psychol. **20**(5), 913–935 (2008)
8. Coskun, B., Herley, C.: Can "something you know" be saved? In: Proceedings of the 11th international conference on information security, pp. 421–440 (2008)
9. Cowan, N.: The magical number 4 in short-term memory: a reconsideration of mental storage capacity. Behav. Brain Sci. **24**(1), 87–114 (2001)
10. Craik, F.I., McDowd, J.M.: Age differences in recall and recognition. J. Exp. Psychol. **13**(3), 474–479 (1987)

11. Dunphy, P., Heiner, A.P., Asokan, N.: A closer look at recognition-based graphical passwords on mobile devices. In: Proceedings of the sixth symposium on usable privacy and security, pp. 3:1–3:12 (2010)
12. Fisher, D.L.: Central capacity limits in consistent mapping, visual search tasks: four channels or more? Cogn. Psychol. **16**(4), 449–484 (1984)
13. Golle, P., Wagner, D.: Cryptanalysis of a cognitive authentication scheme (extended abstract). In: Proceedings of the 2007 IEEE symposium on security and privacy, pp. 66–70 (2007)
14. Hogan, R.M., Kintsch, W.: Differential effects of study and test trials on long-term recognition and recall. J. Verbal Learn. Verbal Behav. **10**(5), 562–567 (1971)
15. Hopper, N.J., Blum, M.: Secure human identification protocols. In: Proceedings of the 7th international conference on the theory and application of cryptology and information security: advances in cryptology, pp. 52–66 (2001)
16. Horowitz, T.S., Wolfe, J.M.: Search for multiple targets: remember the targets, forget the search. Percept. Psychophys. **63**(2), 272–285 (2001)
17. Lei, M., Xiao, Y., Vrbsky, S.V., Li, C.-C., Liu, L.: A virtual password scheme to protect passwords. In: Proceedings of IEEE international conference on communications, pp. 1536–1540 (2008)
18. Li, S., Asghar, H.J., Pieprzyk, J., Sadeghi, A.-R., Schmitz, R., Wang, H.: On the security of PAS (predicate-based authentication service). In: Proceedings of the 2009 annual computer security applications conference, pp. 209–218 (2009)
19. Li, S., Khayam, S.A., Sadeghi, A.-R., Schmitz, R.: Breaking randomized linear generation functions based virtual password system. In: Proceedings of the 2010 IEEE international conference on communications, pp. 23–27 (2010)
20. Li, S., Shum, H.-Y.: Secure human-computer identification (interface) systems against peeping attacks: SecHCI. In: Cryptology ePrint Archive, Report 2005/268 (2005)
21. Long, J., Wiles, J.: No Tech Hacking: A Guide to Social Engineering, Dumpster Diving, and Shoulder Surfing. Syngress, Rockland (2008)
22. Matsumoto, T., Imai, H.: Human identification through insecure channel. In: Proceedings of the 10th annual international conference on theory and application of cryptographic techniques, pp. 409–421 (1991)
23. Nobel, P.A., Shiffrin, R.M.: Retrieval processes in recognition and cued recall. J. Exp. Psychol. **27**(2), 384–413 (2001)
24. Rohrer, D., Wixted, J.T.: An analysis of latency and interresponse time in free recall. Mem. Cogn. **22**(5), 511–524 (1994)
25. Roth, V., Richter, K., Freidinger, R.: A PIN-entry method resilient against shoulder surfing. In: Proceedings of the 11th ACM conference on computer and communications security, pp. 236–245 (2004)
26. Sasamoto, H., Christin, N., Hayashi, E.: Undercover: authentication usable in front of prying eyes. In: Proceeding of the twenty-sixth annual SIGCHI conference on human factors in computing systems, pp. 183–192 (2008)
27. Sternberg, S.: Memory-scanning: mental processes revealed by reaction-time experiments. Am. Sci. **57**, 421–457 (1969)
28. Unsworth, N., Engle, R.W.: The nature of individual differences in working memory capacity: active Maintenance in primary memory and controlled search from secondary memory. Psychol. Rev. **114**(1), 104–132 (2007)
29. Vogel, E.K., Machizawa, M.G.: Neural activity predicts individual differences in visual working memory capacity. Nature **428**(6984), 748–751 (2004)
30. Weinshall, D.: Cognitive authentication schemes safe against spyware (short paper). In: Proceedings of the 2006 IEEE symposium on security and privacy, pp. 295–300 (2006)
31. Wiedenbeck, S., Waters, J., Sobrado, L., Birget, J.-C.: Design and evaluation of a shoulder-surfing resistant graphical password scheme. In: Proceedings of the working conference on advanced visual interfaces, pp. 177–184 (2006)
32. Woodman, G.F., Chun, M.M.: The role of working memory and long-term memory in visual search. Vis. Cogn. **14**(4–8), 808–830 (2006)

33. Woodman, G.F., Luck, S.J.: Visual search is slowed when visuospatial working memory is occupied. Psychon. Bull. Rev. **11**(2), 269–274 (2004)
34. Yan, Q., Han, J., Li, Y., Deng, R.H.: On limitations of designing usable leakage-resilient password systems: attacks, principles and usability. In: Proceedings of the 19th network & distributed system security symposium (NDSS) (2012)
35. Zhao, H., Li, X.: S3PAS: a scalable shoulder-surfing resistant textual-graphical password authentication scheme. In: Proceedings of the 21st international conference on advanced information networking and applications workshops, vol. 02, pp. 467–472 (2007)

Chapter 2
CoverPad: A Leakage Resilient Password System on Touch-Screen Mobile Devices

2.1 Introduction

Mobile devices are widely used in modern life. The password leakage threat to user authentication during password entry is more serious on mobile devices than on desktop computers since mobile devices are often used in public places. While most of prior research on leakage resilient password (LRP) systems [4, 12, 17, 22, 24, 25, 31, 35, 36] focuses on desktop computers, specific restrictions on mobile devices are rarely considered in the design of LRP systems. One restriction is that a mobile device usually has a smaller screen size than a desktop computer. Another restriction is that a mobile device needs to be operable in non-stationary environments such as on public transit. On the other hand, mobile devices provide additional features such as touch screen, which may not be available in traditional settings.

Chapter 1 concludes that a highly secure LRP system without a secure channel inevitably incurs either high memory demand or high cognitive workload which makes it impractical for unaided human users. In this chapter, we introduce a concise yet effective authentication scheme named CoverPad, which is designed for password entry on touch-screen mobile devices [38, 39]. CoverPad improves the leakage resilience of password entry while it retains most of the benefits of legacy passwords. It leverages the gesture detection feature of touch screen in forming a *cover* for user inputs. This cover serves as a temporary secure channel between user and his/her mobile device during password entry, which breaks the correlation between the user's password and an adversary's observations. Due to the use of a temporary secure channel, CoverPad achieves a high level of usability which otherwise cannot be achieved. The high usability is achieved by involving only intuitive cognitive operations and requiring no extra hardware in the design of CoverPad.

Three variants of CoverPad are evaluated in a rigorous user study with realistic testing conditions including *time pressure*, *distraction*, and *mental workload*.

© The Author(s) 2015   29
Y. Li et al., *Leakage Resilient Password Systems*, SpringerBriefs
in Computer Science, DOI 10.1007/978-3-319-17503-4_2

These testing conditions simulate the common use of LRP systems on mobile devices. The user study is conducted to show the influence of these testing conditions on user performance based on some previous work in psychology [11, 19, 21].

2.2 Threat Model

We use the strong passive adversary model as clarified in Sect. 1.2.1, where a passive adversary without any prior knowledge is allowed to record the complete interaction between unaided human *user* and trusted mobile device *server* except for some hidden messages delivered in a temporary secure channel. We focus on the password eavesdropping attacks which exploit the leakage channels *outside* a touch-screen mobile device, including *vision*-based eavesdropping such as hidden camera, *haptics*-based eavesdropping such as physical key logger, and *acoustics*-based eavesdropping such as tone analysis. Figure 2.1 illustrates the attack scenarios used in this chapter. Compared to similar attacks on desktop computers, an adversary in this chapter has more opportunities to launch the password eavesdropping attacks on mobile devices, as mobile devices are widely used in public places. In a crowded area, for example, an adversary may observe password entry in a close distance without being noticed.

In a vision-based attack, an adversary may infer a user's password by observing the movement of the user's fingers. This attack can be performed even without any direct line-of-sight on the screen of the user's mobile device. The adversary's capability can be significantly enhanced with emerging augmented-reality accessories such as Google Glass [16], which is a pair of glasses transferring real-time video captured by a tiny camera to a server and displaying the analyzed results received from the server to the adversary.

The haptics-based attacks are most likely to happen when users use public mobile devices. Mobile devices, such as iPad, have been used as public computer kiosks as observed in museums, restaurants, and hotels [18, 20, 40]. In addition, many existing kiosks are equipped with touch screens. An adversary may install a physical "touch"

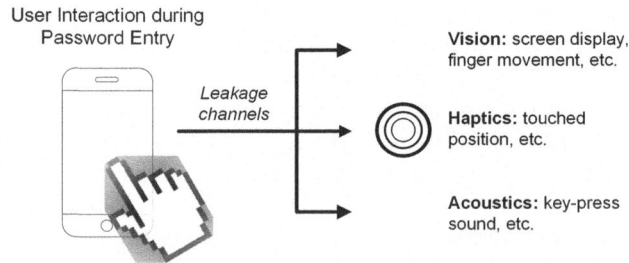

Fig. 2.1 Attack scenarios

logger for launching the haptics-based attacks to passwords. Although such touch logger has not been observed in the wild, it is technically feasible to implement as other physical key loggers [33].

The effectiveness of the acoustics-based attacks depends on whether password entry actions can be distinguished by tone patterns. For example, different tones are played when a user dials different numbers on an old-style phone. Due to environmental noises, the acoustics-based attacks may not be as effective as the vision-based attacks and the haptics-based attacks.

Note that we do not address any password attacks which exploit the leakage channels inside a device, or between the device and a remote server. Such password attacks include *logic key logger*, *malware*, and *network eavesdropping*, which are common to all password-based user authentication schemes. Like most prior research on LRP systems [6, 7, 12, 14, 22, 24, 31], we do not address these attacks in this book. There exist solutions [2, 5, 28, 34] such as application sandbox which can effectively defend against these attacks, though it takes time for them to replace the vulnerable legacy systems. The solutions to address these attacks are independent of password entry and they can be adapted to any user authentication schemes.

Also note that we do not address any active password leakage attacks such as social engineering and phishing [26]. Although their mitigation technologies such as secure URL checker and spam filter have been widely deployed in modern computer systems, some of these attacks cannot be fully addressed using technical solutions alone. Another active attack is the database reading attack, where the back-end database of passwords is hacked and compromised. Since these attacks are orthogonal to the password entry problem, they are out of the scope of this book.

2.3 CoverPad Design

2.3.1 Design Objectives

CoverPad is designed to improve the leakage resilience of password entry while retaining most of the benefits of legacy passwords. The objectives of CoverPad design are described as follows.

In terms of security, CoverPad should minimize the password leakage during password entry in a realistic setting. To achieve this objective, a user should input obfuscated responses derived from his/her password, and/or input his/her password with a secure channel. As it is shown in Chap. 1, it is infeasible for unaided users to use obfuscated responses due to the limitation of human cognitive capabilities. Therefore, it is necessary to rely on certain secure channel to achieve this security objective. Since it is usually difficult, even impractical, to protect all messages delivered between user and server, a hybrid solution is chosen in the design of CoverPad, where a temporary secure channel is conveniently established during password entry for simple obfuscation of passwords.

In the presence of a temporary secure channel, it is possible to achieve the optimal security objective—*no password leakage*. As long as the temporary secure channel is not compromised, CoverPad provides the same leakage resilience as *one-time pad* [29], where the most efficient attacks for an adversary to learn about the password are online dictionary attacks. The security analysis of CoverPad is provided in the next section.

In terms of usability, CoverPad should preserve most of the benefits of legacy passwords in order to be practical [9]. The major benefits of legacy passwords include no extra devices being required, and only intuitive cognitive operations being performed. On mobile devices, additional restrictions include relatively small screen size as compared to desktop computers, and non-stationary environment in which CoverPad should operate. CoverPad minimizes the number of visual elements that are displayed simultaneously on a mobile device screen, and simplifies the operations to be performed during password entry.

2.3.2 Conceptual Design

The conceptual design of CoverPad is shown in Fig. 2.2, where the hidden transformation $T_i(\cdot)$ is a random mapping $\Omega \rightarrow \Omega$, where Ω is the set of all individual elements contained in a password alphabet. At the beginning of password entry, a user performs a hand-shielding gesture to view the current hidden transformation T_1 for the first character a_1 in his/her password. Then, the user applies T_1 to a_1 and enters the transformed response e_1. This procedure repeats for each password element a_i. During the whole password entry, T_i disappears immediately once the gesture is not detected. The user can always view T_i by performing the gesture again before inputting e_i.

Setup:
A server and a user agree on a k-length password pwd $= (a_1, a_2, \ldots, a_k)$, *where a* **password element** $a_i = pwd[i]$ *belongs to an alphabet with size w. It is allowed that* $a_i = a_j$, *for* $i \neq j$.

Password Entry:
For each i from $[1,k]$:

Step 1: *The touch screen shows a keypad with all the elements in the alphabet.*

Step 2: *The user is asked to perform a hand-shielding gesture to read the hidden transformation* $T_i(\cdot)$ *protected by the hand-shielding gesture.* $T_i(\cdot)$ *will immediately disappear if the gesture is no longer detected.*

Step 3: *The user clicks on response element* e_i, *where* $e_i = T_i(a_i) = (a_i + r_i \mod w)$, *where* r_i *is a random number drawn from a uniform distribution. A new random number* r_i *is generated for each round i. The hand-shielding gesture is not required for this step.*

Fig. 2.2 Conceptual design of CoverPad

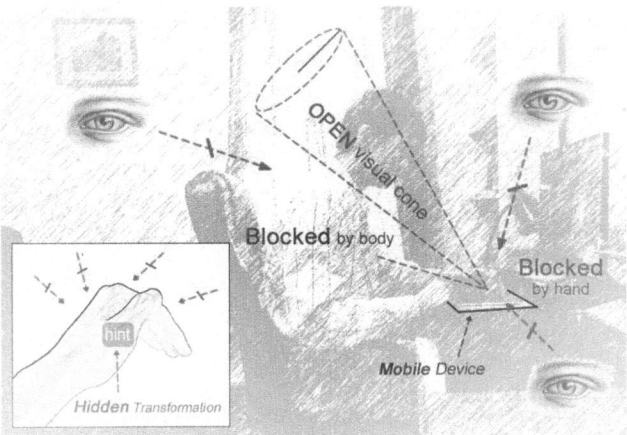

Fig. 2.3 Hand-shielding gesture and its effectiveness

Figure 2.3 illustrates how to correctly perform a hand-shielding gesture in CoverPad. This gesture restricts the vision channel to a small visual cone. The visual cone is not accessible to an adversary unless the adversary's eyes are close enough to the user's head, which makes the adversary easily exposed. A hidden camera near the line of sight may help capture the hidden transformation. However, it needs to be adjusted according to the user's height and the current position of mobile device, which may lead to the user's awareness. On the other hand, the observable responses for the same password element are uniformly randomized. Thus, CoverPad is immune to the haptics-based eavesdropping.

It is difficult to compromise the temporary secure channel formed by the hand-shielding gesture in external eavesdropping attacks, though the use of this gesture is simple. Once the protective gesture is not detected by the touch screen, the hidden transformation disappears such that the hidden transformation is always protected under the required gesture. Note that a hidden transformation alone does not leak any information about the password. As long as the hidden transformation is not revealed together with the corresponding response, an adversary cannot infer any valuable information about the actual password. A proof about this property is given in Sect. 2.4.

2.3.3 Implementation Variants

Figure 2.4 illustrates three variants of CoverPad, including NumPad-Add, NumPad-Shift, and LetterPad-Shift, which are tailored for users with different skill sets.

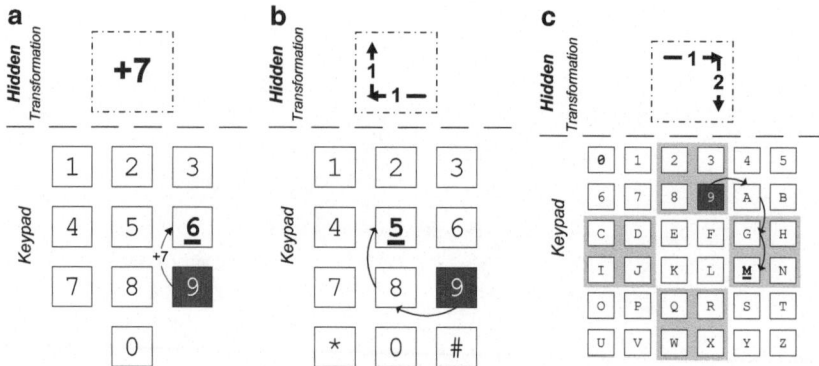

Fig. 2.4 Three variants of CoverPad. (**a**) NumPad-Add, (**b**) NumPad-Shift, (**c**) LetterPad-Shift

2.3.3.1   NumPad-Add

In NumPad-Add, the alphabet of password consists of digits 0–9 only. The hidden transformation is performed by adding a random digit to the current password element and then performing "mod 10" if the sum is larger than 9, where the value of the random digit ranges from 0 to 9. For example, the correct response in the first round is $6 = (9 + 7) \mod 10$ given password 934567 and the hidden message "*plus 7.*"

2.3.3.2   NumPad-Shift

In NumPad-Shift, the alphabet of password consists of digits 0–9 only. The hidden transformation is performed by *shifting* the location of the current password element by X-offset and Y-offset, where the offset values are randomly taken from $\{-1, 0, 1\}$ for X-offset, and $\{-1, 0, 1, 2\}$ for Y-offset. For a 3×4 keypad design shown in Fig. 2.4b, the transformed response for a_i is calculated as $pad[x(a_i) + \Delta x \mod 3][y(a_i) + \Delta y \mod 4]$, where Δx is the X-offset, Δy is the Y-offset, $x(a_i)$ is the X-index of a_i, and $y(a_i)$ is the Y-index of a_i. For example, the correct response in the first round is 5 if the password is 934567 and the hidden message is "*move left by 1 step and move up by 1 step.*"

Note that two extra keys * and # are added to the keypad; otherwise, the distribution of hidden transformations is not uniform on the keypad layout. The proof for the necessity of these two keys is given as follows. Assuming * and # keys are removed, the keypad now contains only 10 keys for digits 0–9. To provide a full transformation from a secret key to a random key, the minimum value set is $\{-1, 0, 1\}$ for X-offsets and $\{-1, 0, 1, 2\}$ for Y-offsets. There are twelve combinations between X-offsets and Y-offsets, but only ten keys on the keypad. If the offset values are drawn from a uniform distribution, certain response keys for a given password element would have a higher frequency compared to others (it is similar as placing twelve balls in ten buckets in a deterministic way). The exact distribution of response keys is decided by the underlying password element, thus it discloses valuable information about the password. Similarly, if the response keys are drawn from a uniform distribution, the offset values are not uniformly distributed. Therefore, it is necessary to add these two extra keys to the NumPad-Shift keypad.

2.3.3.3 LetterPad-Shift

In LetterPad-Shift, the alphabet of password consists of letters a to z and digits 0–9 (36 elements in total). The hidden transformation is the same as NumPad-Shift. The offset values are randomly taken from $\{-2, -1, 0, 1, 2, 3\}$ for both X-offset and Y-offset for a 6×6 keypad design. The transformed response for a_i is calculated as $pad[x(a_i) + \Delta x \mod 6][y(a_i) + \Delta y \mod 6]$. As shown in Fig. 2.4c, a background grid can be added on the keypad so as to facilitate the calculation.

2.4 Security Analysis

We focus on the passive password attacks which exploit vision, haptics, or acoustics channels between users and mobile devices as clarified in Sect. 2.2. For CoverPad, a strong passive adversary in these attacks can observe all user's actions and inputs, while the hidden transformation is protected by a temporary secure channel. From the user's inputs, the adversary knows that the i-th pressed key is derived from the i-th element in the user's password. However, the adversary cannot further infer what the i-th password element is, as proved below.

Given a pressed key e_i, and two password elements a_x and a_y in a w-sized password alphabet, let $Pr(a_x|e_i)$ and $Pr(a_y|e_i)$ be the probabilities for e_i being pressed while the underlying password element is a_x and a_y, respectively. We have $Pr(a_x|e_i) = Pr(e_i = a_x + r_i \mod w) = Pr(r_i = e_i - a_x \mod w) = Pr(r_i = C \mod w) = 1/w = Pr(a_y|e_i)$ for any meaningful i, x, and y, where C is a constant integer randomly drawn from a uniform distribution. Therefore, a sequence of pressed keys observed by an adversary is equivalent to a random sequence, which completes the proof.

CoverPad achieves no password leakage provided that the hidden transformation is protected by the hand-shielding gesture. As long as the hidden transformation is not disclosed together with the corresponding response, an adversary cannot infer any information about the underlying password except for the length of password even after an infinite number of observations.

2.5 Usability Evaluation

A rigorous user study is conducted to evaluate the usability of CoverPad.

2.5.1 Methodology

The participants in the user study are recruited from undergraduate students in Singapore Management University. There are 61 participants in total, 30 male and 31 female, with age range between 20 and 25. These participants come from five different schools, in which 42 of them have a social science or business related background, and the remaining have a computer science or information technology related background. Each participant is paid with 10 dollars as compensation for their time. A ranking system is set up so that each participant can see a *performance score* representing how well he/she performs compared to other participants. This ranking system provides a bit motivation for the participants to do their best in the user study. A numerical identifier is assigned to each participant in order to protect user privacy.

The user study is conducted in a quiet room, where each participant is asked to use all three variants of CoverPad in three *test groups*, one variant per teat group. These variants are implemented on Apple iPad, which are referred to as *schemes* in this section. The order of the schemes is randomized to avoid the learning effect that affects the performance for a specific scheme. For each test group, a user is required to memorize a *randomly generated* password in the beginning. The password strength is set to be equivalent to 6-digit PIN, where the password length is four for LetterPad-Shift, and six for both NumPad-Shift and NumPad-Add. The same password is used for the same test group; a "*show my password*" button is provided in case a participant forgets his/her password during the user study. The participants learn how to use a scheme in an interactive step-by-step tutorial. The participants are required to go through the whole tutorial for the first scheme appearing in the test, and they may skip the tutorial for the second and third schemes after learning the basic scheme design. At the end of each tutorial, there is a short pretest for the participants to exercise. If a participant fails to pass the pretest, an on-site help is provided so as to ensure that the participant understands how to use the scheme before the user test starts.

In each test group, there are six tests simulating various conditions in which CoverPad is in use, including the influence of time pressure, distraction, and mental workload. The order of these tests is randomized so as to avoid any learning effect.

There are total 18 tests in three test groups. Since the participants may feel exhausted or bored in a long test, each test is designed to be short, which can be finished within 1 or 2 min. Also, the participants are given a short break after each test group. At the end of the user study, the participants are given a questionnaire for collecting their perceptions on the schemes. The whole user study takes 35–50 min to complete.

2.5.2   Test Conditions

In order to simulate various conditions in which CoverPad is in use, two experimental tools, including timer and secondary task, are introduced in the user study. A timer is used to create a time pressure by showing a participant how much time is left for the current test. It is implemented as a progress bar whose length increases every second with a countdown text field showing how many seconds are left. In addition, certain CRT (*choice reaction time*) tasks, which are commonly used in experimental psychology [11, 19, 21], are used as secondary tasks for simulating the unexpected distraction and persistent mental workload. The CRT tasks occupy the central executive in human brain [3] and require participants to give distinct responses for each possible stimulus. In the user study of CoverPad, the participants perform the CRT tasks by pressing the correct button among N buttons, where the correct button has the same color as the stimulus. For example, if the stimulus shows a red button, a participant should press the red button among N buttons with different colors. For the tests in the distraction condition, $N = 2$ is chosen since the major focus is to unexpectedly disrupt the password entry with a CRT task. For the tests in the mental workload condition, $N = 8$ is used to create a considerable mental workload, which is the same as in the classic Jensen Box setting [21].

In the user study of CoverPad, six different conditions are formed for each test group by combining two modes and three statuses. The two modes w.r.t. timer are given below:

- *Relaxed mode:* A participant is asked to minimize the error rate in a fixed number of login attempts while time is not considered in the calculation of performance score. The number of login attempts is 5 for no-extra-task status and 3 for distraction status and mental workload status.
- *Timed mode:* A participant is asked to perform as many successful logins as possible within 1 min while both time and accuracy are considered in the calculation of performance score. The countdown of a timer creates a time pressure during login.

The three statuses w.r.t. secondary tasks are given below:

- *No-extra-task status:* A participant is asked to perform the login task only.
- *Distraction status:* A simple CRT task may appear with 1/3 probability each time when a participant presses a response key. This task creates unexpected distractions during password entry.
- *Mental workload status:* A relatively complex CRT task appears each time when a participant presses a response key. This task creates a continuing mental workload during password entry.

Among the six conditions, the combination of *relaxed* mode and *no-extra-task* status is referred to as the *normal condition*, which is the common condition usually tested in most prior work on LRP systems [4, 6, 7, 12, 14, 17, 22, 24, 25, 31, 35, 36]. For ease of reference, Table 2.1 gives the short names for the test conditions.

Table 2.1 Short name of test condition

Short name	Test condition
Normal	*Relaxed* mode + *no-extra-task* status
Timed	*Timed* mode + *no-extra-task* status
Distraction	*Relaxed* mode + *distraction* status
Distraction + timed	*Timed* mode + *distraction* status
Mental workload	*Relaxed* mode + *mental workload* status
Mental workload + timed	*Timed* mode + *mental workload* status

The following hypotheses about the test conditions are used in the user study:

(H1)   Compared to the normal condition, the login time is significantly shorter when a time pressure is present.

(H2)   Compared to the normal condition, the login accuracy is significantly lower when a time pressure is present.

(H3)   Compared to the normal condition, the login time is significantly longer when an unexpected distraction is present.

(H4)   Compared to the normal condition, the login accuracy is significantly lower when an unexpected distraction is present.

(H5)   Compared to the normal condition, the login time is significantly longer when a persistent mental workload is present.

(H6)   Compared to the normal condition, the login accuracy is significantly lower when a persistent mental workload is present.

(H7)   Compared to a condition in the relaxed mode with secondary tasks, the login time is significantly shorter for its counterpart in the timed mode.

(H8)   Compared to a condition in the relaxed mode with secondary tasks, the login accuracy is significantly lower for its counterpart in the timed mode.

2.5.3 Learning Curve

CoverPad requires a different process for password entry compared to legacy passwords. Even with tutorial and pretests, it is observed that some participants are impatient to read all instructions and keep pressing the next button. Figure 2.5 compares the user performance under the normal condition when the CoverPad variants are arranged in different test groups. It shows that the user performance is significantly worse in the first test groups. The differences between the second group and the third group are less significant. The learning curve implies that most participants get familiar with CoverPad after the first test groups. Therefore, the first test groups are considered as part of the learning process; only the performance data collected from the second and third test groups are used in the following user study.

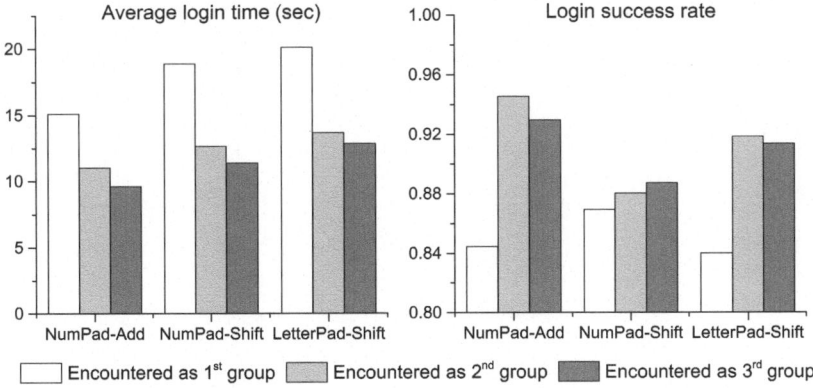

Fig. 2.5 Learning curve of CoverPad

2.5.4 Experimental Results

In testing, the participants' performance is measured using the following metrics: average login time, login success rate, round success rate, and average edit distance. A *round* success rate is the average success rate for a user to correctly input one password element after applying a hidden transformation. An *edit distance* is the minimum number of insertions, deletions, substitutions, and adjacent transpositions required to transform an input string into the correct password string; an *average* edit distance is the average value of edit distances calculated from all login attempts under certain test condition. Among these metrics, the login success rate, the round success rate, and the average edit distance are used to evaluate *login accuracy*.

Some statistical tools are used to test the significance of experimental results, where a significance level of $\alpha = 0.05$ is used. For each comparison, an *omnibus* test is run across all test conditions for each scheme. Since all performance data are quantitative, the *Kruskal-Wallis* (KW) test is used for omnibus tests, which is an analogue of ANOVA but does not require normality. If the omnibus test indicates significance, the *Mann-Whitney* (MW) U test is further used to perform pair-wise comparisons so as to identify specific pairs with significant differences. The statistical test results are given in the Appendix of this chapter.

2.5.4.1 Performance Under Normal Condition

In the normal condition, a participant is asked to perform login tasks without any time pressure or secondary tasks. The normal condition corresponds to the combination of the relaxed mode and the no-extra-task status; it is used as a *baseline* in testing CoverPad.

Figure 2.6a shows the average time for a successful login attempt in the normal condition. For all three schemes, most participants completed their logins within

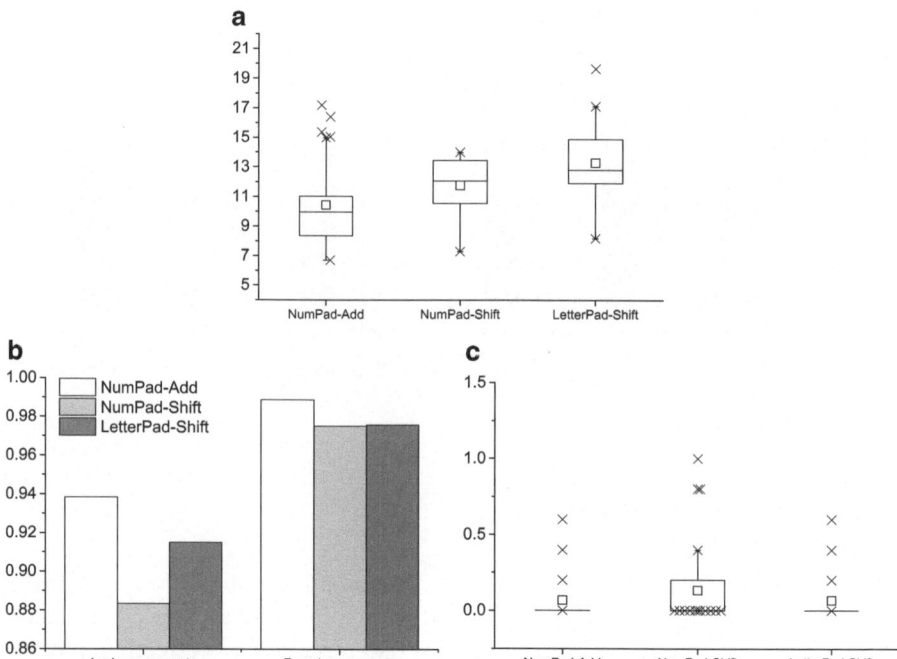

Fig. 2.6 Average login time, success rate, and edit distance under normal condition. (**a**) Login time distribution (s), (**b**) average success rate, (**c**) edit distance distribution

13 s. Figure 2.6b, c show the login accuracy in this case. The number of login attempts is restricted to five; therefore, a single mistake would lower the login success rate down to 80 %. Most participants made *at most* one mistake when they used CoverPad schemes for the first time after a short training. This is shown by 97.5 % average round success rate and 0.13 average edit distance in the worst case. Particularly, for the distribution of average edit distance of NumPad-Shift, 27 participants among 40 samples generated an average edit distance of zero (i.e. no mistakes during all tests), which are shown as a cluster of *outliers* at the bottom of the box chart.

2.5.4.2   Influence of Time Pressure

Figure 2.7 shows the impact of time pressure without any secondary tasks. The results show that the participants behaved much hastily in the presence of time pressure. The average time for a successful login attempt becomes shorter; the login accuracy decreases as well. The statistical test results given in the Appendix show significant difference w.r.t the login time ($p = 0.017$ for NumPad-Add and $p < 0.001$ for LetterPad-Shift) but no difference w.r.t. the login accuracy. Therefore, the hypothesis *H1* is supported while *H2* is not.

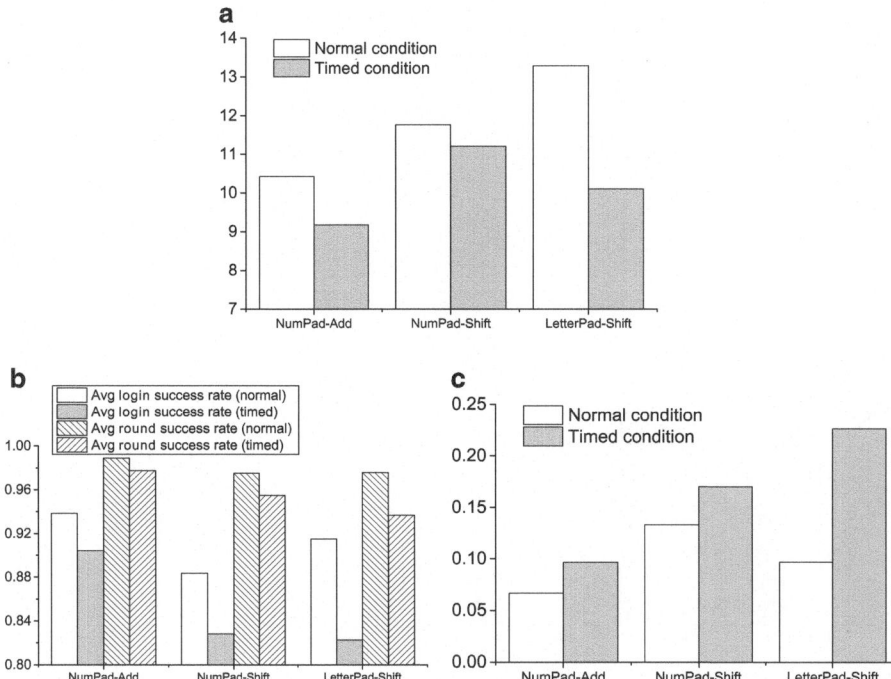

Fig. 2.7 Impact of time pressure. (**a**) Average login time (s), (**b**) average success rate, (**c**) average edit distance

The insignificant results w.r.t. the login accuracy are mainly due to the *ceiling* effect [1]; that is, the tests are not difficult enough to distinguish the influence of different test conditions. An evidence of this effect is that the majority of the participants made no mistakes during all the tests, which is shown in the Appendix.

2.5.4.3 Influence of Distraction

Figure 2.8 shows the impact of distraction without any time pressure. Many participants made a mistake when they saw a distraction task for the first time (however, NumPad-Shift is an exception). For NumPad-Add and LetterPad-Shift shown in Fig. 2.8b, the round success rate returns to a comparable level as in the normal condition after the first time when the distraction task appears. This is reasonable since the distraction task is no longer a surprise for the participants. However, even after the participants get familiar with the distraction tasks, compared to the normal condition, the success rate is still lower, the average edit distance is larger, and the average login time is longer. Nonetheless, the statistical test results (see the Appendix) show that the differences are not significant. Therefore, neither hypothesis *H3* nor *H4* is supported.

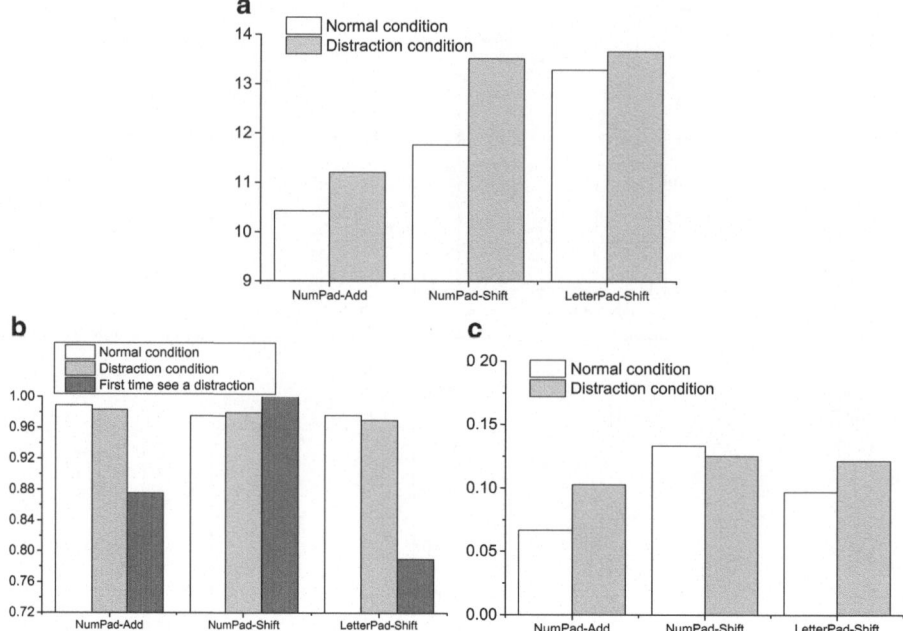

Fig. 2.8 Impact of distraction. (**a**) Average login time (s), (**b**) average **round** success rate, (**c**) average edit distance

2.5.4.4   Influence of Mental Workload

Figure 2.9 shows the impact of mental workload without any time pressure. The average login time becomes significantly longer with some mental workload ($p = 0.003$ for NumPad-Add) due to the context switch in a user's mind between password inputs and secondary CRT tasks. An extra startup time is required to release the central executive after each CRT task. This is simulated when users cannot get rid of other thoughts during password entry. The actual effect of mental workload depends on the status of a user's mind. The impact may be elevated when the actual mental workload is higher than the CRT tasks used in user study. On the other hand, the login accuracy is lower than that in the normal condition; the difference is however not significant due to the ceiling effect. Therefore, hypothesis *H5* is supported but *H6* is not. Persistent mental workload slows the process of password entry significantly for CoverPad.

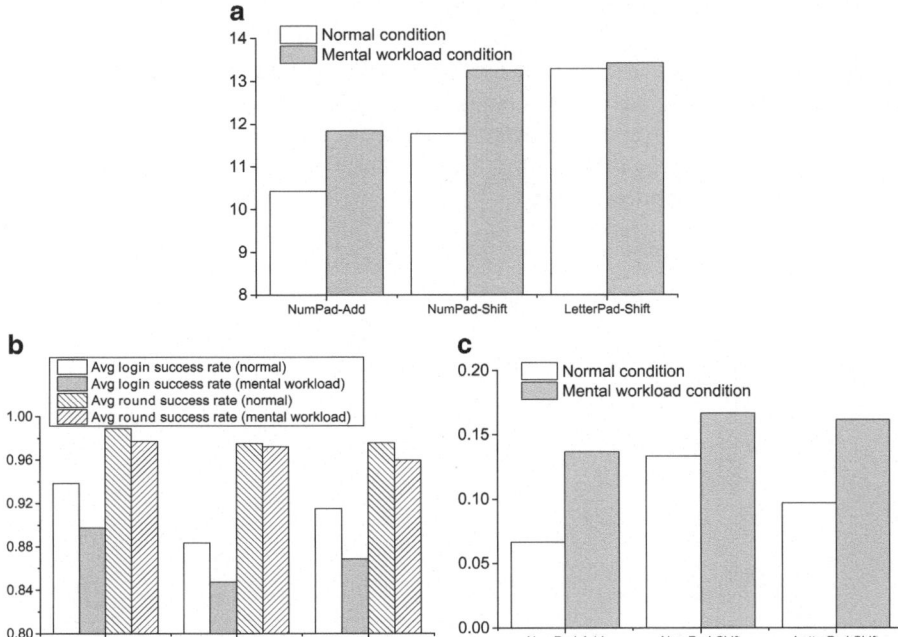

Fig. 2.9 Impact of mental workload. (**a**) Average login time (s), (**b**) average success rate, (**c**) average edit distance

2.5.4.5 Performance Under Combined Conditions

The overall impact of combined conditions is examined when the distraction or mental workload appears together with the time pressure. As expected, compared to the cases without time pressure, the average login time becomes shorter (from 11.7 to 10.3 s on average), the login success rate becomes lower (from 87.5 to 81.3 %), and the average edit distance becomes larger (from 0.151 to 0.243). The statistical test results given in the Appendix show that the differences in terms of login time are significant ($p = 0.009$ for NumPad-Add, $p = 0.019$ for NumPad-Shift, and $p < 0.001$ for LetterPad-Shift) and the differences in terms of login accuracy are not significant due to the ceiling effect. Therefore, hypothesis *H7* is supported but *H8* is not. The time pressure is an effective stimulus to speed up password entry even in the presence of secondary tasks.

2.5.4.6   Effectiveness of Secondary Tasks

Figure 2.10 shows the accuracy of performing the secondary tasks, which is the percentage of the secondary tasks being correctly performed by each participant under certain test condition. The overall average accuracy is 98.3 % across all test conditions. The accuracy is high; thus, the CRT tasks work as intended in disturbing users' mind during password entry.

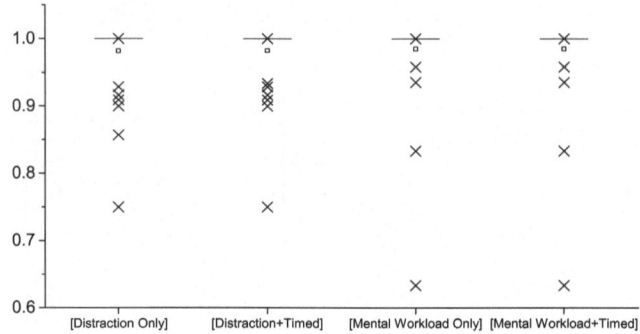

Fig. 2.10 Accuracy of performing secondary tasks

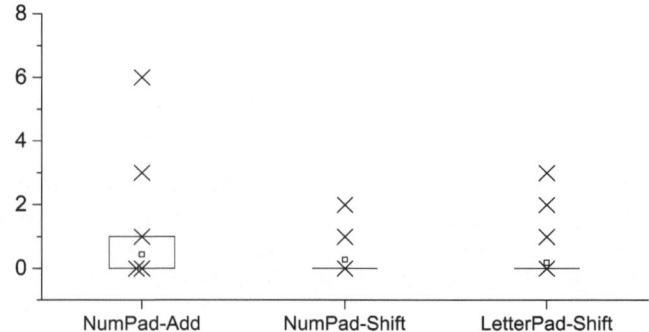

Fig. 2.11 Total number of times for pressing "*show my password*" button per participant

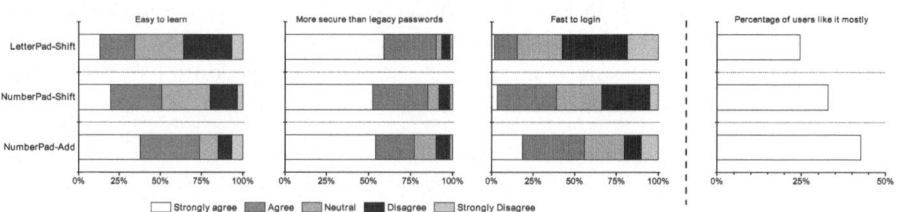

Fig. 2.12 Perception of participants

2.5.4.7   Memory Interference by Mental Calculation

Figure 2.11 shows how frequently a participant presses the "*show my password*" button during all tests in a test group. Note that the participants are not allowed to write down their assigned passwords, but they can always click that button in case they forgot their passwords. The overall average value for this frequency is 0.31 across all three test groups. The mental calculation involved in the hidden transformation of CoverPad does not pose a significant interference on participants' capability of recalling their passwords.

2.5.4.8   User Perception

Figure 2.12 shows the perception of participants collected from questionnaires. The results indicate that the participants generally feel that CoverPad schemes are secure and easy to use. While NumPad-Add is the most popular, the other two schemes also have their favorite users.

2.5.5   Comparison with Legacy Passwords

Table 2.2 gives a comparison between CoverPad and legacy password based on the *usability-deployability-security* metrics proposed in [9], where a metric is not shown if neither CoverPad schemes nor legacy password offers the corresponding benefit. In the comparison, CoverPad schemes are rated as immature since they have not been widely deployed. Coverpad schemes are not *server-compatible*, as most current servers support static and replayable passwords only, which could be changed in the future. CoverPad schemes are *quasi-resilient-to-internal-observation* in a sense that any key logger or malware which fails to capture the hidden transformation causes no password leakage. Overall, CoverPad schemes significantly improve the security strength while retaining most of the benefits provided by legacy password.

2.6   Discussions

2.6.1   Device Screen Size

CoverPad can be implemented on devices with large screens such as Apple iPad; it can also implemented on devices with small screens such as Apple iPhone, which is illustrated in Fig. 2.13. In the latter case, a user may use one hand A to perform the hand-shielding gesture, and use the other hand B to hold the phone. The thumb on hand B can be used to press the response keys. If implemented on mobile phones

Table 2.2 Comparison between CoverPad and legacy password (● means "offer benefit," ○ means "almost offer benefit," and empty means "offer no benefit")

	Nothing-to-carry	Easy-to-learn	Efficient-to-use	Infrequent-errors	Easy-recovery-from-loss	Accessible	Negligible-cost-per-user	Server-compatible	Browser-compatible	Mature	Non-proprietary	Resilient-to-physical-observation	Resilient-to-targeted-impersonation	Resilient-to-internal-observation	Resilient-to-theft	No-trusted-third-party	Requiring-explicit-consent	Unlinkable
CoverPad schemes	●	●	○	○	●	●	●		●		●	●	○	○	●	●	●	●
Legacy password	●	●	●	○	●	●	●	●	●	●	●			○	●	●	●	●

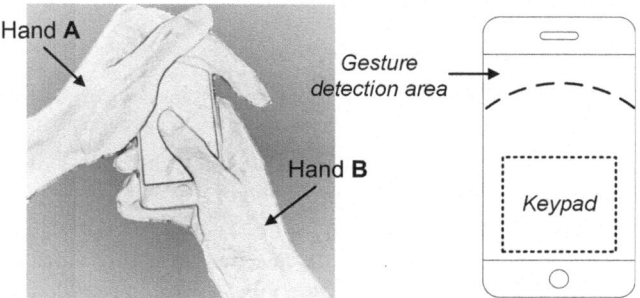

Fig. 2.13 CoverPad on small screen

with larger touch screens such as Samsumg Galaxy Note II, a user may not be able to click all the keys with the thumb of hand *B* that holds the device. To deal with this situation, he/she may use one hand *A* to perform the hand-shielding gesture and key pressing sequentially. Once the user raises his/her hand before pressing a key, the hidden transformation immediately disappears because the gesture is no longer detected by the touch screen. Meantime, the user does not need to worry about whether the actual keys being pressed or the finger movements during key pressing may be observed by an adversary, as the sequence of the pressed keys alone does not leak any information about the underlying password as analyzed in Sect. 2.4.

CoverPad can be implemented on any devices equipped with touch screens. For example, many ATM machines have been deployed with touch screens. CoverPad can be implemented on such machines so as to mitigate the threat of the ATM skimming attack [23].

2.6.2 Limitations

Ecological validity is a challenging issue in any user study. Like most prior research [14, 17, 22, 24, 25], the user experiments with CoverPad engage only university students. These participants are younger and more educated than general population. The usability evaluation may vary with other populations. The user study is also restricted by the sample size, which may affect the results of statistical tests. Typical examples are the insignificant results on the login accuracy of CoverPad schemes. Moreover, the user study does not include any experiments on memory effects. Since the same alphabet and password composition are used in CoverPad as in legacy password systems, the impact of memory effects on the user performance would be similar to legacy password systems as shown in the literature [13, 32].

2.7 Bibliographic Note

Although the design of LRP systems has been investigated for more than two decades [27], it remains a challenge to achieve both leakage resilience and high usability at the same time. Almost all previous LRP schemes for unaided users with acceptable usability have been broken [4, 25, 35, 36]. Strong evidences show that it is necessary to rely on certain secure channels between users and LRP systems so as to achieve high security and usability [10, 37]. A few schemes [6, 7, 12, 14, 22, 24, 31] were designed in this direction. Among them, CoverPad is mostly inspired by the concept of physical metaphor introduced in [22]. CoverPad distinguishes itself from prior works in the sense that it not only achieves leakage resilience but also retains most benefits of legacy password systems, while some of prior schemes are flawed in terms of security [14, 31], and the others incur extra usability costs. The extra usability costs required by other LRP systems include (i) using uncommon devices such as gaze tracker [12, 24], haptic motor [7], and large pressure-sensitive screen [22], (ii) requiring accessory devices [6], and (iii) being inoperative in a non-stationary environment [7].

The idea of applying random transformations to a fixed password for password entry is not new in the design of LRP systems; the challenge is to make it *human-friendly*. CoverPad addresses this challenge by exploiting a temporary secure channel which can be easily established on devices with touch screens. It is thus easier to use than some existing LRP and patented schemes. Taking GridCode [15] as an example, it requires users to memorize extra secrets (besides passwords) and perform certain calculations in order to perform the transformations specified in its scheme design. Another advantage of CoverPad is that each character of the password uses a different hidden transformation; thus, a single character would be exposed if a transformation is exposed in CoverPad. In comparision, GridCode uses the same transformation for all the characters in a password. If the hidden transformation in GridCode is disclosed, the entire password would be exposed.

People have proposed generic principles for design LRP schemes. Roth et al. [30] proposed using a cognitive trapdoor game to transform the knowledge of an underlying password into obfuscated responses. Li and Shum [25] later suggested three other principles including time-variant responses, randomness in challenges and responses, and indistinguishability against statistical analysis. Yan et al. [37] further extended the coverage by including the design principles against the brute force attacks and statistical attacks. The design of CoverPad is consistent with these design principles.

Bonneau et al. [9] proposed a generic framework for evaluating user authentication proposals which emphasizes the importance of retaining the benefits of legacy password systems. The framework introduces twenty-five benefits covering usability, deployability, and security. Other research work on password-based user authentication can be found in a survey paper [8].

2.8 Conclusion

CoverPad, a leakage-resilient password entry scheme leveraging the touch screen feature of mobile devices, is introduced in this chapter. CoverPad achieves leakage resilience while preserves most of the benefits of legacy password systems. The usability of CoverPad is evaluated in a user study in various test conditions related to time pressure, distraction, and mental workload. Among these conditions, the time pressure and mental workload are shown to have significant impacts on users' performance.

Appendix: Statistical Test Results

The statistical test results for login time shown in Table 2.3 indicates that the same test condition may have difference impacts on the login time with different schemes, where the statistically significant results are marked with ★.

The statistical test results on login accuracy are not shown since none of them are significant. This is due to the ceiling effect, as it is shown in Table 2.4, where each cell in this table shows the percentage of the participants who make no mistakes under certain test condition. In the worst case, 50.0 % participants make no mistakes in all tests, which implies that the tests are not difficult enough to distinguish various test conditions with respect to the login accuracy. The CoverPad variants are easy to use even in the presence of time pressure, distraction, and mental workload. Note that this does not necessarily mean that these factors will not influence the login accuracy of other user authentication schemes.

Table 2.3 The statistical test results for login time (s)

Average login time of NumPad-Add - *omnibus* KW $\chi_5^2 = 32.423$, $p < 0.001$		
Normal (10.4)	Timed (9.2)	U = 551, $p = 0.017$ ★
	Distraction (11.2)	U = 679, $p = 0.184$
	Distraction + timed (10.3)	U = 878, $p = 0.989$
	Mental workload (11.8)	U = 515, $p = 0.003$ ★
	Mental workload + timed (10.7)	U = 696, $p = 0.319$
Distraction (11.2)	Distraction + timed (10.3)	U = 718, $p = 0.107$
Mental workload (11.8)	Mental workload + timed (10.7)	U = 558, $p = 0.009$ ★

Average login time of NumPad-Shift - *omnibus* KW $\chi_5^2 = 11.965$, $p = 0.034$		
Normal (11.7)	Timed (11.2)	U = 666, $p = 0.199$
	Distraction (13.5)	U = 645, $p = 0.137$
	Distraction + timed (11.7)	U = 727, $p = 0.485$
	Mental workload (13.3)	U = 655, $p = 0.164$
	Mental workload + timed (11.4)	U = 644, $p = 0.135$
Distraction (13.5)	Distraction + timed (11.7)	U = 565, $p = 0.024$ ★
Mental workload (13.3)	Mental workload + timed (11.4)	U = 555, $p = 0.019$ ★

Average login time of LetterPad-Shift - *omnibus* KW $\chi_5^2 = 49.252$, $p < 0.001$		
Normal (13.2)	Timed (10.1)	U = 294, $p < 0.001$ ★
	Distraction (13.6)	U = 774, $p = 0.667$
	Distraction + timed (11.0)	U = 413, $p < 0.001$ ★
	Mental workload (13.4)	U = 653, $p = 0.116$
	Mental workload + timed (11.5)	U = 472, $p = 0.002$ ★
Distraction (13.6)	Distraction + timed (11.0)	U = 422, $p < 0.001$ ★
Mental workload (13.4)	Mental workload + timed (11.5)	U = 631, $p = 0.075$

Table 2.4 Evidence of ceiling effect in statistical tests on login accuracy

	NumPad-Add (%)	NumPad-Shift (%)	LetterPad-Shift (%)
Normal	82.9	67.5	75.6
Timed	78.0	62.5	53.7
Distraction	80.5	70.0	63.4
Distraction + timed	70.7	55.0	58.5
Mental workload	75.6	57.5	65.9
Mental workload + timed	65.9	50.0	51.2

References

1. Ceiling effect. http://en.wikipedia.org/wiki/Ceiling_effect
2. Apple. Mac os x. http://www.apple.com/osx/
3. Baddeley, A.D., Hitch, G.: Working memory. Psychol. Learn. Motiv. **8**, 47–89 (1974)

4. Bai, X., Gu, W., Chellappan, S., Wang, X., Xuan, D., Ma, B.: PAS: predicate-based authentication services against powerful passive adversaries. In: Proceedings of the 2008 annual computer security applications conference, pp. 433–442 (2008)
5. Begemann, O.: Remote View Controllers in iOS 6. http://oleb.net/blog/2012/10/remote-view-controllers-in-ios-6
6. Bianchi, A., Oakley, I., Kostakos, V., Kwon, D.S.: The phone lock: audio and haptic shoulder-surfing resistant pin entry methods for mobile devices. In: Proceedings of the 5th international conference on tangible, embedded, and embodied interaction, pp. 197–200 (2011)
7. Bianchi, A., Oakley, I., Kostakos, V., Kwon, D.S.: Obfuscating authentication through haptics, sound and light. In: Proceedings of the 2011 annual conference on human factors in computing systems, pp. 1105–1110 (2011)
8. Biddle, R., Chiasson, S., van Oorschot, P.C.: Graphical passwords: learning from the first twelve years. ACM Comput. Surv. **44**(4), 19 (2012)
9. Bonneau, J., Herley, C., van Oorschot, P.C., Stajano, F.: The quest to replace passwords: a framework for comparative evaluation of web authentication schemes. In: Proceedings of IEEE symposium on security and privacy (2012)
10. Coskun, B., Herley, C.: Can "something you know" be saved? In: Proceedings of the 11th international conference on information security, pp. 421–440 (2008)
11. Craik, F.I., McDowd, J.M.: Age differences in recall and recognition. J. Exp. Psychol. **13**(3), 474–479 (1987)
12. De Luca, A., Denzel, M., Hussmann, H.: Look into my eyes!: Can you guess my password? In: Proceedings of the 5th symposium on usable privacy and security, pp. 7:1–7:12 (2009)
13. De Luca, A., Denzel, M., Hussmann, H.: Towards understanding ATM security: a field study of real world ATM use. In: Proceedings of the sixth symposium on usable privacy and security (2010)
14. De Luca, A., von Zezschwitz, E., Husmann, H.: Vibrapass: secure authentication based on shared lies. In: Proceedings of the 27th international conference on human factors in computing systems, pp. 913–916 (2009)
15. Ginzburg, L., Sitar, P., Flanagin, G.K.: User authentication system and method. US Patent 7,725,712, SyferLock Technology Corporation (2010)
16. Google. Google Glass. http://plus.google.com/+projectglass
17. Hopper, N.J., Blum, M.: Secure human identification protocols. In: Proceedings of the 7th international conference on the theory and application of cryptology and information security: advances in cryptology, pp. 52–66 (2001)
18. Hotel, H.B.: iPAD—Free for Every Hotel Guest. http://www.hollmann-beletage.at/en/ipad
19. Imbo, I., Vandierendonck, A.: The role of phonological and executive working memory resources in simple arithmetic strategies. Eur. J. Cogn. Psychol. **19**(6), 910–933 (2007)
20. Imran, A.: iPADs can now be used as public kiosks. http://www.redmondpie.com/ipad-public-kiosks-video/
21. Jensen, A.R.: Process differences and individual differences in some cognitive tasks. Intelligence **11**(2), 107–136 (1987)
22. Kim, D., Dunphy, P., Briggs, P., Hook, J., Nicholson, J.W., Nicholson, J., Olivier, P.: Multi-touch authentication on tabletops. In: Proceedings of the 28th international conference on human factors in computing systems, pp. 1093–1102 (2010)
23. Krebs. Would You Have Spotted the Fraud? http://krebsonsecurity.com/2010/01/would-you-have-spotted-the-fraud
24. Kumar, M., Garfinkel, T., Boneh, D., Winograd, T.: Reducing shoulder-surfing by using gaze-based password entry. In: Proceedings of the 3rd symposium on usable privacy and security, pp. 13–19 (2007)
25. Li, S., Shum, H.-Y.: Secure human-computer identification (interface) systems against peeping attacks: SecHCI. In: Cryptology ePrint Archive, Report 2005/268 (2005)
26. Long, J., Wiles, J.: No Tech Hacking: A Guide to Social Engineering, Dumpster Diving, and Shoulder Surfing. Syngress, Rockland (2008)

27. Matsumoto, T., Imai, H.: Human identification through insecure channel. In: Proceedings of the 10th annual international conference on theory and application of cryptographic techniques, pp. 409–421 (1991)
28. Microsoft. Windows 8. http://windows.microsoft.com
29. Miller, F.: Telegraphic Code to Insure Privacy and Secrecy in the Transmission of Telegrams. C.M. Cornwell, New York (1882)
30. Roth, V., Richter, K., Freidinger, R.: A PIN-entry method resilient against shoulder surfing. In: Proceedings of the 11th ACM conference on computer and communications security, pp. 236–245 (2004)
31. Sasamoto, H., Christin, N., Hayashi, E.: Undercover: authentication usable in front of prying eyes. In: Proceeding of the 26th annual SIGCHI conference on human factors in computing systems, pp. 183–192 (2008)
32. Shay, R., Kelley, P.G., Komanduri, S., Mazurek, M.L., Ur, B., Vidas, T., Bauer, L., Christin, N., Cranor, L.F.: Correct horse battery staple: exploring the usability of system-assigned passphrases. In: Proceedings of the eighth symposium on usable privacy and security (2012)
33. Spycop. Hardware Keylogger Detection. http://spycop.com/keyloggerremoval.htm
34. TCG. Trusted Computing Group. http://www.trustedcomputinggroup.org
35. Weinshall, D.: Cognitive authentication schemes safe against spyware (short paper). In: Proceedings of the 2006 IEEE symposium on security and privacy, pp. 295–300 (2006)
36. Wiedenbeck, S., Waters, J., Sobrado, L., Birget, J.-C.: Design and evaluation of a shoulder-surfing resistant graphical password scheme. In: Proceedings of the working conference on advanced visual interfaces, pp. 177–184 (2006)
37. Yan, Q., Han, J., Li, Y., Deng, R.H.: On limitations of designing leakage-resilient password systems: attacks, principles and usability. In: Proceedings of the 19th annual network and distributed system security symposium (2012)
38. Yan, Q., Han, J., Li, Y., Zhou, J., Deng, R.H.: Designing leakage-resilient password entry on touchscreen mobile devices. In: Proceedings of the 8th ACM symposium on information, computer and communications security (ASIACCS), pp. 37–48 (2013)
39. Yan, Q., Han, J., Li, Y., Zhou, J., Deng, R.H.: Leakage-resilient password entry: challenges, design, and evaluation. Comput. Secur. **48**(2015), 196–211 (2015)
40. ZDNet. More iPAD Love: Now Hotels Offer iPAD to Customers. http://www.zdnet.com/blog/apple/more-ipad-love-now-hotels-offer-ipad-to-customers/6850

Chapter 3
ShadowKey: A Practical Leakage Resilient Password System

3.1 Introduction

Mobile devices are increasingly widely used for connecting people to the cyberspace. Users may use smartphones, tablets, or wearable smart devices to login to their personal accounts such as internet banking and social network accounts, and access to various services such as financial, health, and corporate services. In order to prevent unauthorized access to such accounts and services, user authentication is required to verify user identity. Password based authentication is still the most pervasive authentication solution in practice due to its benefits over other authentication solutions such as smartcards and biometrics [1]. However, password-based user authentication suffers from an intrinsic weakness of password leakage during password entry. This threat is more serious when user authentication is performed on mobile devices and in public places.

Designing leakage-resilient password (LRP) systems for unaided users remains a challenge despite two decades of intensive research. Most LRP systems were broken soon after their proposals, while the remnants are very difficult to use. In Chap. 1, the fundamental limitations on the general design aspects of LRP systems are addressed. It is demonstrated that most of the existing LRP systems are subject to brute force attacks and statistical attacks. To thwart these attacks, a set of design principles are identified to achieve leakage resilience in the absence of any secure channel. It is also shown that these attacks cannot be effectively mitigated without sacrificing the usability of LRP systems in the absence of any secure channels between users and LRP systems. An LRP system with no secure channel always imposes a considerable amount of cognitive workload on its users, which indicates that a highly usable LRP system must incorporate certain secure channel in design.

Chapter 2 introduces CoverPad, which leverages a temporary secure channel between user and touch-screen mobile device to achieve leakage resilience for password entry. CoverPad requires a user to perform certain transformation operations in his/her mind so as to input each transformed password symbol on a touch screen.

Y. Li et al., *Leakage Resilient Password Systems*, SpringerBriefs in Computer Science, DOI 10.1007/978-3-319-17503-4_3

Such transformation operations, though much simpler than the cognitive operations required in other LRP systems, still impose a considerable amount of cognitive workload to users.

In this chapter, we introduce a new LRP system, which is named as ShadowKey [3]. ShadowKey leverages either permanent secure channel, which naturally exists between user and the display unit of certain types of mobile devices such as smart glasses, or temporary secure channel, which can be easily realized between user and touch screen by placing a hand-shielding gesture, to achieve better usability in leakage resilient password entry. A significant advantage of ShadowKey in comparison with other LRP systems is that it does not require users to perform any cognitive operations beyond remembering the original passwords during password entry. Most conveniently, a user may use voice to input a password without leaking any information about the password (except for the length of the password) to the surroundings. Alternatively, a user may use a touch screen to input his/her password while a hand-shielding gesture is removed without password leakage. ShadowKey is suitable to be implemented on a wide variety of devices such as smart glasses, smart phones, laptops, and touch-screen desk-tops and ATMs. With ShadowKey, users can input their passwords/PINs easily in public places without leaking out any information about their passwords during password entry.

3.2   Setting of ShadowKey

ShadowKey is an LRP system which enables a user to input his/her password into a mobile device without leaking any information about the password, except for the length of the password, to strong passive adversaries. The setting of ShadowKey is illustrated in Fig. 3.1. The LRP system consists of a processing unit connecting to a display unit, an input unit, a memory, and an output unit. It is assumed that all units of ShadowKey and the communication channels between them work in a secure environment without being accessed by any malwares (in other words, how to provide such secure environment is independent to the focus of this chapter). There may exist a secure channel between the display unit and the user's eyes, while the channel between the input unit and the user is open to potential attackers.

We use the strong passive adversary model as used in the previous two chapters, where a passive adversary without any prior knowledge is allowed to record the complete interaction between unaided human user and the input unit. We focus on the password eavesdropping attacks which exploit the leakage channels outside ShadowKey, including vision-based eavesdropping such as hidden camera, haptics-based eavesdropping such as physical key logger, and acoustics-based eavesdropping such as tone analysis. Mobile devices are more vulnerable to such attacks compared to traditional desktop computers since mobile devices are widely used in public places where adversaries may observe password entry without being noticed.

Fig. 3.1 Setting of ShadowKey

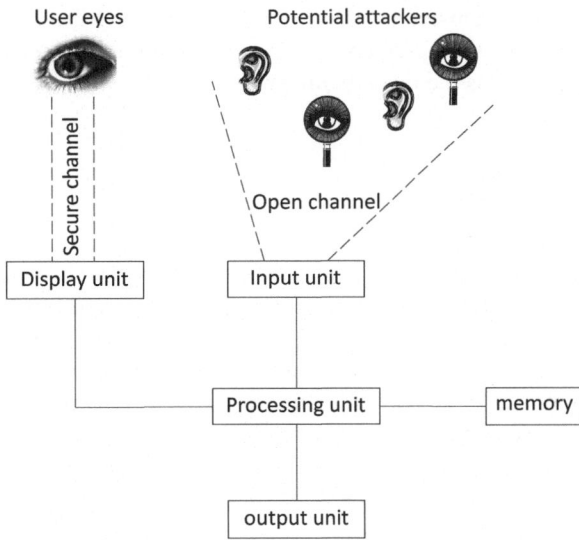

3.2.1 Secure Channel

A secure channel between the user's eyes and the display unit ensures that the information displayed on the display unit is viewed by the legitimate user only. In other words, potential attackers surrounding the user cannot discern the information on the display unit when a secure channel is present.

A secure channel can be either permanent or temporary. A permanent secure channel exists between the user and the display unit of certain types wearable devices such as smart glasses. A temporary secure channel exists between the user and a touch-screen while a hand-shielding gesture is detected on the touch screen. In the latter case, the secure channel disappears if the user removes his/her hand-shielding gesture from the touch screen. As it is explained in Chap. 2, a temporary secure channel is easy to form and it is effective to restrict a passive adversary from accessing the information displayed on the touch screen.

3.2.2 Input Unit

The input unit is used by a user to input his/her password or other information in the presence of adversaries who monitor the input process. The input unit can be either voice input unit, or touch-screen input unit. The voice input unit takes the user's voice as input and converts the voice signals into meaningful information, while the touch-screen input unit shows certain keypads on its display, which can be pressed or selected by a user during the input process. Note that the information displayed on the touch-screen input unit is also accessible to any adversary. It is possible that the same touch screen is used as the input unit (when no secure channel is detected) and also as the display unit (when a secure channel is detected).

3.2.3 Processing Unit

The processing unit is used to run a series of operations for leakage resilient password entry, taking a user's input from the input unit, sending information to the display unit, storing programs and data in the memory, and outputting the user's password to the output unit. The output unit could be any user authentication application such as unlocking a device, activating a password manager, accessing a user account, or making a payment.

3.2.4 Keypad, Shadow Pad, and Display Unit

A *keypad* is a specific arrangement of symbols, including all symbols which are used to form a password. A keypad may be associated with one or more shadow pads, where a *shadow pad* consists of the same set of symbols as the keypad, but these symbols are randomly permuted from their original positions in the keypad. Here, "random permutation" means that for any shadow pad symbol e_i and any keypad symbols a_x and a_y, the probability $Pr(a_x|e_i)$ for e_i being the input while the underlying password symbol is a_x is the same as the probability $Pr(a_y|e_i)$ for e_i being the input while the underlying password symbol is a_y.

If the keypad is too big to fit into the display unit, it may be partitioned into multiple smaller keypads so that each small keypad can be shown on the display unit. In such case, a shadow pad is partitioned in the same way as the keypad into several small shadow pads.

When a secure channel is present during password entry, both keypad and shadow pad appear on the display unit. Figure 3.2 shows a keypad of 10 digits as well as a shadow pad displayed on the display unit. As long as the secure channel is removed, the shadow pad disappears, and only the keypad remains on the display unit.

Fig. 3.2 Key pad and shadow pad

3.3 Design of ShadowKey

In this section, we present the design of ShadowKey. First, we assume that the whole keypad as well as an associated shadow pad can be shown together on the display unit. Later, we present two solutions to address the case in which the keypad is too big to fit into the display unit.

3.3.1 ShadowKey with One Keypad

First, consider the case where the keypad and an associated shadow pad fit in the display unit. When a user views both keypad and shadow pad on the display unit via a secure channel, he/she may locate a password symbol in the keypad, and input the associated symbol from the shadow pad on the input unit. After the user inputs the current symbol and before he/she inputs another one, a new (random) shadow pad is generated to replace the existing shadow pad. Figure 3.3 illustrates the process of inputting a password on the input unit.

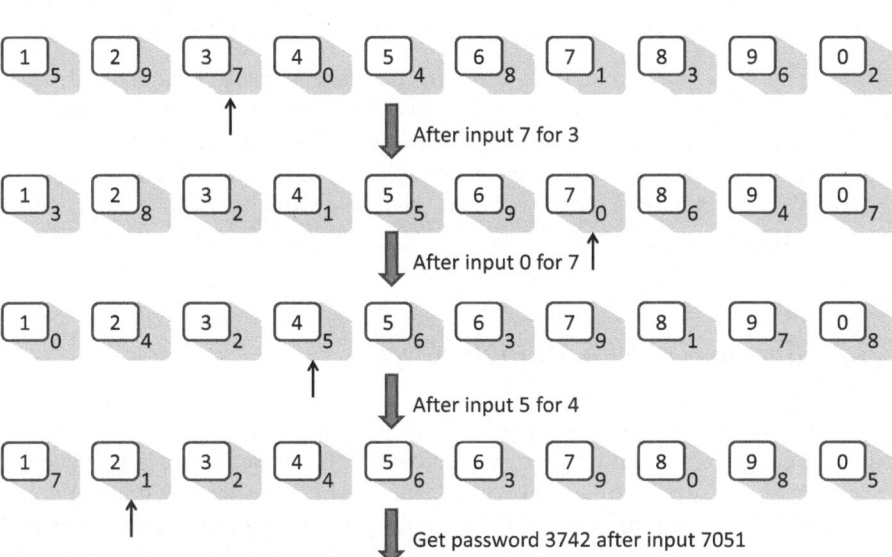

Fig. 3.3 Inputting password with ShadowKey

The process of password entry with ShadowKey is given below:

1. The user starts a new process of password entry (e.g., by inputting "start" in the input unit).
2. The processing unit sets $n = 1$; then, the processing unit generates the keypad, stores it in the memory, and displays it on the display unit.
3. The processing unit generates a new shadow pad for the keypad, stores it in the memory, and displays it together with the keypad on the display unit when a secure channel is detected.
4. The user on the display unit may locate the n-th password symbol in the keypad, and input its associated symbol from the shadow pad on the input unit.
5. If no input is received for a certain period of time, or the user inputs "cancel" at any time, then the processing unit clears the memory and stops the process.
6. Upon receiving a shadow pad symbol from the input unit, the processing unit in the memory locates the symbol on the shadow pad, retrieves the associated password symbol from the keypad, and stores the symbol as the n-th password symbol.
7. The user may choose to input on the input unit (i) "retry:" go to step 3; (ii) "finish:" the processing unit sends all retrieved password symbols from the memory to the output unit, clears the memory, and stops the process; (iii) "cancel:" the processing unit clears the memory and stops the process.
8. If no input is received for a certain period of time, or the user inputs "continue" at any time, the processing unit increases n by one and goes to step 3.

ShadowKey is most convenient to use with permanent secure channel and voice input unit. It is also easy to use in the case of temporary secure channel and voice input unit, and in the case of permanent secure channel and touch-screen input unit. In the case of temporary secure channel and touch-screen input unit, the user may use the same touch screen as the display unit (when a secure channel is detected) and as the input unit (when the secure channel is removed).

3.3.2   ShadowKey with Multiple Keypads

Now, consider the case in which a keypad is too big to fit into the display unit. The processing unit may partition the keypad into multiple smaller keypads {keypad[1], ..., keypad[m]} such that keypad=$\cup_i$keypad[i], where the intersections of some small keypads may not be empty. The processing unit also partitions a shadow pad (which is randomly permuted from the whole keypad) into multiple smaller shadow pads in the same way so that each small keypad and its associated small shadow pad can be displayed together on the display unit.

Figure 3.4 illustrates one example of multiple keypads which are used by some smart phones including iPhones. In this example, the keypad on top is used to input both lower case and upper case English letters. For each English letter shown in the keypad, it is mapped to two shadow pad symbols, one for the upper case letter, and

Fig. 3.4   Example of multiple keypads

the other for the lower case letter. Figure 3.5a illustrates the layout of such letter 'A' and its shadow pad symbols, where 'D' corresponds to upper case 'A,' and '2' corresponds to lower case 'a.' Figure 3.5b illustrates symbol '0' on a keypad which is mapped to only one shadow pad symbol '@.'

Fig. 3.5   (**a**) Shadow pad symbol 'D' for upper case 'A' and shadow pad symbol '2' for lower case 'a'; (**b**) Shadow pad symbol '@' for '0'

The steps 2–4 in the process of password entry with ShadowKey given in Sect. 3.3.1 should be revised as following in the case of multiple keypads:

2. The processing unit sets $n = 1$; then, it generates the whole keypad, and partitions it into m small keypads forming a keypad pool = {keypad[1],...,keypad[m]}; then, it stores all keypads in the memory
3. The processing unit selects a keypad X randomly from the keypad pool and displays it on the display unit. Then, it generates a new shadow pad for the entire keypad, partitions it into m small shadow pads the same way as it partitions the keypad, stores all the shadow pads in the memory, and displays the shadow pad corresponding to keypad X together with X on the display unit when a secure channel is detected.
4. If the user locates the n-th password symbol on the current keypad, he or she may input the associated symbol of the shadow pad on the input unit; else the user may input "next" on the input unit, in which case the processing unit removes X from the keypad pool (if the resulting keypad pool is empty, then it is set to {keypad[1],...,keypad[m]}−{X}) and goes to step 3.

In the above step 4, if the user inputs a shadow pad symbol on a touch screen input unit, the user should select the shadow pad symbol on a keypad displayed on the touch screen (after the user removes a hand-shielding gesture). To facilitate this input, the processing unit displays keypad[1] first; if the user cannot find the shadow pad symbol on this keypad, he or she may input "next" and the processing unit displays keypad[2], and so on (keypad [1] is shown after keypad[m]). Note that the keypads are displayed in a sequential order until the user discovers the shadow pad symbol on a keypad and inputs it successfully.

3.3.3   ShadowKey with One Customized Keypad

Still, consider the case in which a keypad is too big to fit into the display unit. To avoid the inconvenience of surfing among multiple keypads during password entry, the user may choose to customize a keypad which includes all the user's password symbols such that the customized keypad and its shadow pad can fit into the display unit. A shadow pad in this case consists of the shadow pad symbols associated with the customized keypad symbols, where the shadow pad symbols are taken from the whole shadow pad, which is a random permutation of the whole keypad.

The process of password entry with one customized keypad is similar to the process with one keypad as given in Sect. 3.3.1 except that if a secure channel is detected on the display unit, the customized keypad and a shadow pad are displayed; otherwise, the original keypad is partitioned into a number of small keypads and they are displayed in a sequential order on the display unit or on a touch-screen input unit.

It is convenient to use the customized keypad with a voice input unit. As it is shown below, ShadowKey is leakage resilient even if the customized keypad is leaked to an adversary. Note that the leakage of the customized keypad would however weaken the strength of a user's password against other attacks such as brute-force attacks and dictionary attacks (which are out of the scope of this book).

3.4   Analysis of ShadowKey

3.4.1   Security

For ShadowKey, a strong passive adversary without any prior knowledge may record the complete interaction between an unaided human user and the input unit, while the mapping between keypads and shadow pads is protected by a secure channel. From the user's inputs, the adversary knows that the i-th input symbol is taken from a shadow pad and that the symbol is associated with the i-th symbol in the user's password. However, the adversary cannot further infer what the i-th password symbol is, as proved below.

Given any input symbol e_i from a shadow pad, and any password symbol a_x from a w-sized keypad, let $Pr(a_x|e_i)$ be the probability for e_i being the input while the underlying password symbol is a_x. In the case of ShadowKey with one keypad, we have $Pr(a_x|e_i) = \frac{1}{w}$.

Now, consider the case of multiple keypads $\{keypad[1], \ldots, keypad[m]\}$ which sizes are $w_1, \ldots, w_m$, respectively. Let $shadow[j]$ denote the shadow pad associated with $keypad[j]$ for $j = 1, \ldots, m$. If a_x belongs to one keypad only, say $a_x \in keypad[k]$ without loss of generality, we have

$$Pr(a_x|e_i) = Pr(a_x \in keypad[k]|e_i)$$

$$= Pr(a_x \in keypad[k]|e_i \in shadow[1]) \cdot Pr(e_i \in shadow[1]) + \ldots$$

$$+ Pr(a_x \in keypad[k]|e_i \in shadow[k]) \cdot Pr(e_i \in shadow[k]) + \ldots$$

$$+ Pr(a_x \in keypad[k]|e_i \in shadow[m]) \cdot Pr(e_i \in shadow[m])$$

$$= 0 \cdot \frac{w_1}{w} + \ldots + \frac{1}{w_k} \cdot \frac{w_k}{w} + \ldots + 0 \cdot \frac{w_m}{w}$$

$$= \frac{1}{w}$$

If a_x belongs to ℓ ($\ell \leq m$) multiple keypads due to possible overlaps of these keypads in some keypad designs, it is easy to derive $Pr(a_x|e_i) = \frac{\ell}{w}$.

In the case of one customized keypad which size is $w' < w$, we have $Pr(a_x|e_i) = \frac{1}{w}$ if the adversary does not know the composition of the customized keypad beyond its size w'; otherwise, the adversary may increase the probability to $Pr(a_x|e_i) = \frac{1}{w'}$.

In all cases, given a password symbol a_x, the probability $Pr(a_x|e_i)$ remains unchanged no matter which e_i is observed by an adversary, and how many e_i's are observed. In this sense, ShadowKey achieves no password leakage. An adversary would infer no information about a password except its length even after an infinite number of observations provided that the shadow pads are protected by a secure channel.

3.4.2 Usability

It is concluded in Chap. 1 that unaided human may not be competent to use a secure LRP system without a secure channel due to the high cognitive workload or memory demand for using such LRP system. In Chap. 2, CoverPad is proposed to achieve leakage resilience while preserving most of the benefits of legacy password systems. Nonetheless, CoverPad still incurs considerable mental calculation for performing certain addition, modular, or shifting operations; it also incurs a considerable memory demand for remembering a hidden transformation which is required for inputting a password symbol.

ShadowKey further improves the usability by eliminating any mental calculation or memory demand during password entry. With ShadowKey, a user does not need to perform any mental calculation beyond recognizing each password symbol on a keypad and inputting the associated shadow pad symbol on the input unit. In addition, a user does not need to remember anything else except the original password during password entry. ShadowKey is most convenient to use with a permanent secure channel (as provided by smart glasses) and a voice-input channel (as provided by most mobile devices); in such case, a user can directly speak out the shadow pad symbols which are associated with password symbols for password entry. In other cases, a temporary secure channel is used to protect the mappings between keypads and shadow pads. In all cases, the usability of ShadowKey is clearly better than what is achieved in CoverPad.

3.5 Bibliographic Note

ShadowKey relies on either permanent secure channel or temporary secure channel to achieve a high usability for leakage resilient password entry. ShadowKey does not require users to perform any mental calculations or remember anything beyond the original passwords during password entry. In comparison, many existing LRP systems incur a considerably higher amount of cognitive workload.

Due to the negative result about the LRP systems without any secure channels [6], CoverPad makes use of a temporary secure channel between user and touch-screen mobile device to achieve no leakage for password entry [7, 8]. CoverPad requires a user to perform certain transformation operations in his/her mind so as to input a transformed password symbol on a touch screen. Such transformation operations, though much simpler than what is required in previous LRP systems [6], still impose a non-negligible amount of cognitive workload to users.

In particular, at the beginning of password entry in CoverPad, a user needs to perform a hand-shielding gesture and view a fresh "hidden transformation" T for the first symbol in his/her password. Then, the user applies T to the first password symbol and enters the transformed symbol. This procedure repeats for each password symbol. During the whole password entry, T disappears immediately once the gesture is not being detected. A user can always view T by performing the gesture again before inputting a transformed symbol.

CoverPad can be implemented in three variants, including NumPad-Add, NumPad-Shift, and LetterPad-Shift. In NumPad-Add, the alphabet of password consists of digits 0–9 only. The hidden transformation is performed by adding a random digit to the current password element and then performing "mod 10," where the value of the random digit ranges from 0 to 9.

In NumPad-Shift, the alphabet of password also consists of digits 0–9 only. The keypad is arranged in an array of four rows and three columns with two additional symbols '⋆' and '#.' The hidden transformation is performed by shifting the location of the current password symbol by X-offset and Y-offset, where the offset values are randomly taken from $\{-1,0,1\}$ for X-offset, and $\{-1,0,1,2\}$ for Y-offset.

In LetterPad-Shift, the alphabet of password consists of letters 'a' to 'z' and digits 0–9 (36 elements in total). The keypad is arranged in an array of six rows and six columns. The hidden transformation is the same as NumPad-Shift. The offset values are randomly taken from $\{-2,-1,0,1,2,3\}$ for both X-offset and Y-offset.

Though CoverPad preserves most of the benefits of legacy password systems, it still incurs considerable mental calculation for performing addition, modular or shifting operations; it also requires that a user remember a hidden transformation in his/her mind for inputting each password symbol. In comparison, a ShadowKey user does not need to perform any algorithmic operations for password entry. What the user needs to do is to simply localize password symbols on the keypad, and input the associated shadow pad symbols on the input unit. Another advantage of ShadowKey is that it is most convenient to use with voice input, especially when a permanent secure channel is present.

In [2], Lev Ginzburg et al. used a similar transformation to map a user's password to a one-time password so as to achieve leakage resilience in user authentication. In their invention, a user is required to remember a formula (e.g, $(A+B)\cdot I\cdot(g^z)+31$) and calculate the result of the formula using the values assigned by the authentication system to the variables for password entry. In comparison, ShadowKey does not require users to remember any secret formula or perform any formula calculations for password entry. Therefore, it is much easier to use from a user's point of view.

In [5], Hwa-Shik Shin proposed to use a random keypad for inputting passwords. A user may input his/her password directly on a "password keypad" where ten numbers (from 0 to 9) randomly arranged, or on an "indirect password keypad" where the ten numbers are randomly mapped to ten letters (from A to J). Compared to [5], ShadowKey has the following advantages:

- Leakage resilience: The innovation proposed in [5] is not leakage resilient since it does not prevent any attacker who monitors a user's password entry from obtaining the user's password. The problem it addresses is that if an identical password is repetitively input on a fixed keypad, a recognizable trace may remain on the buttons of the keypad corresponding to the password; consequently, an attacker may infer the password or at least reduce the search space for the password by examining the trace after the password entry. In comparison, ShadowKey is leakage resilient even if an attacker monitors the password entry process or examines the trace after the password entry.
- More use cases: ShadowKey is designed specifically to suit for multiple use cases in combinations of (i) permanent secure channel or temporary secure channel, (ii) voice input or touch-screen input, and (iii) one keypad, multiple keypads, or one customized keypad, while the previous innovation [5] does not specify any secure channel, voice input, multiple keypads, or customized keypad in forming use cases.
- Input on fixed keypad: In the case of touch-screen input, ShadowKey enables users to input on fixed keypad(s), while the previous innovation [5] requires users to input on a keypad with randomly arranged numbers or letters.

In [4], McIntyre et al. proposed a method for secure PIN entry on touch-screen display. A plurality of numerical keypad layouts are defined such that each layout has a unique arrangement of decimal number locations. Each PIN entry event is performed on a numerical keypad layout randomly chosen from the defined layouts. The advantages of ShadowKey compared to [4] are similar as compared to [5] except that the PIN entry event in [4] is performed on a numerical keypad layout which is randomly chosen from a set of defined layouts.

3.6 Conclusion

ShadowKey enables unaided users to input their passwords in public places without leaking out any information about their passwords during password entry. The designed LRP system is easy to use in a sense that users do not need to remember anything else except their passwords, and they do not need to perform any transformation operations in their minds as it is required in other LRP systems. ShadowKey protects the mappings between keypads and shadow pads via either permanent secure channels, which naturally exist between users and display units on certain mobile devices such as smart glasses, or temporary secure channels while hand-shielding gestures are detected on touch-screen mobile devices. Most conveniently, users can use voice input for password entry. Alternatively, users may use touch screens for password input under no secure channel. ShadowKey can thus be implemented on a wide variety of devices such as smart glasses, smart phones, laptops, and touch-screen desk-tops and ATMs for users to input their passwords/PINs easily in public places without password leakage.

References

1. Bonneau, J., Herley, C., van Oorschot, P.C., Stajano, F.: The quest to replace passwords: a framework for comparative evaluation of web authentication schemes. In: Proceedings of IEEE symposium on security and privacy (2012)
2. Lev, G., Paul, S., George, K.F.: User authentication system and method. US Patent: US7725712 B2 (2010)
3. Li, Y.: Method for leakage resilient password entry. Singapore Provisional Patent 10201405977P (2014)
4. McIntyre, K.E., Sheets, J.F.: Gougeon, D.A.J., Watson, C.W., Morlang, K.P., Faoro, D.: Method for secure PIN entry on touch screen display. US Patent: US 6549194 B1 (2003)
5. Shin, H.-S.: Device and method for inputting password using random keypad. US Patent: US7698563 B2 (2010)
6. Yan, Q., Han, J., Li, Y., Deng, R.H.: On limitations of designing leakage-resilient password systems: attacks, principles and usability. In: Proceedings of the 19th annual network and distributed system security symposium (2012)
7. Yan, Q., Han, J., Li, Y., Zhou, J., Deng, R.H.: Designing leakage-resilient password entry on touchscreen mobile devices. In: Proceedings of the 8th ACM symposium on information, computer and communications security (ASIACCS), pp. 37–48 (2013)
8. Yan, Q., Han, J., Li, Y., Zhou, J., Deng, R.H.: Leakage-resilient password entry: challenges, design, and evaluation. Comput. Secur. **48**(2015), 196–211 (2015)